METHODS OF STATISTICAL MODEL ESTIMATION

METHODS OF STATISTICAL MODEL ESTIMATION

Joseph M. Hilbe

Jet Propulsion Laboratory
California Institute of Technology, USA
and
Arizona State Univeristy, USA

Andrew P. Robinson

ACERA & Department of Mathematics and Statistics
The University of Melbourne, Australia

CRC Press
Taylor & Francis Group
Boca Raton London New York

CRC Press is an imprint of the
Taylor & Francis Group, an **informa** business

A CHAPMAN & HALL BOOK

CRC Press
Taylor & Francis Group
6000 Broken Sound Parkway NW, Suite 300
Boca Raton, FL 33487-2742

First issued in paperback 2019

ISBN-13: 978-1-4398-5802-8 (hbk)
ISBN-13: 978-0-367-38000-7 (pbk)

Visit the Taylor & Francis Web site at
http://www.taylorandfrancis.com

and the CRC Press Web site at
http://www.crcpress.com

Contents

Preface

Methods of Statistical Model Estimation has been written to develop a particular pragmatic viewpoint of statistical modelling. Our goal has been to try to demonstrate the unity that underpins statistical parameter estimation for a wide range of models. We have sought to represent the techniques and tenets of statistical modelling using executable computer code. Our choice does not preclude the use of explanatory text, equations, or occasional pseudo-code. However, we have written computer code that is motivated by pedagogic considerations first and foremost.

An example is in the development of a single function to compute deviance residuals in Chapter 4. We defer the details to Section 4.7, but mention here that deviance residuals are an important model diagnostic tool for generalized linear models (GLMs). Each distribution in the exponential family has its own deviance residual, defined by the likelihood. Many statistical books will present tables of equations for computing each of these residuals. Rather than develop a unique function for each distribution, we prefer to present a single function that calls the likelihood appropriately itself. This single function replaces five or six, and in so doing, demonstrates the unity that underpins GLM. Of course, the code is less efficient and less stable than a direct representation of the equations would be, but our goal is clarity rather than speed or stability.

This book also provides guidelines to enable statisticians and researchers from across disciplines to more easily program their own statistical models using R. R, more than any other statistical application, is driven by the contributions of researchers who have developed scripts, functions, and complete packages for the use of others in the general research community. At the time of this writing, more than 4,000 packages have been published on the Comprehensive R Archive Network (CRAN) website.

Our approach in this volume is to discuss how to construct several of the foremost types of estimation methods, which can then enable readers to more easily apply such methods to specific models. After first discussing issues related to programming in R, developing random number generators, numerical optimization, and briefly developing packages for publication on CRAN, we discuss in considerable detail the logic of major estimation methods, including ordinary least squares regression, iteratively re-weighted least squares, maximum likelihood estimation, the EM algorithm, and the estimation of model parameters using simulation. In the process we provide a number of guidelines that can be used by programmers, as well as by statisticians and researchers in general regarding statistical modelling.

Datasets and code related to this volume may be found in the *msme* package on CRAN. We also will have R functions and scripts, as well as data, available for download on Prof. Hilbe's BePress Selected Works website, `http://works.bepress.com/joseph_hilbe/`. The code and data, together with errata and a PDF document named `MSME_Extensions.pdf`, will be in the `msme` folder on the site. The extensions document will have additional code or guidelines that we develop after the book's publication that are relevant to the book. These resources will also be available at the publisher's website, `http://www.crcpress.com/product/ISBN/9781439858028`.

Readers will find that some of the functions in the package have not been exported, that is, made explicitly available when the package is loaded in memory. They can still be called, using the protocol shown in the following example for `Sjll`:

```
msme:::Sjll( ... insert arguments here as usual! ...)
```

That is, prepend the library name and three colons to the function call.

We very much encourage feedback from readers, and would like to have your comments regarding added functions, operations and discussions that you would like to see us write about in a future edition. Our goal is to have this book be of use to you for writing your own code for the estimation of various types of models. Moreover, if readers have written code that they wish to share with us and with other readers, we welcome it and can put it in the Extensions ebook for download. We will always acknowledge the author of any code we use or publish. Annotated, self-contained code is always preferred!

Readers are assumed to have a background in R programming, although the level of programming experience necessary to understand the book is rather minimal. We attempt to explain every R construct and operation; therefore, the text should be successfully used by anyone with an interest in programming. Of course the more background one has in using R, and in programming in general, the easier it will be to implement the code we developed for this volume.

Overview

Chapter 1 is the introductory chapter providing readers with the basics of R programming, including the specifics of R objects, functions, matrices, object-oriented programming, and creating R packages.

Chapter 2 deals with the nature of statistical models and of maximum likelihood estimation. We then introduce random number generators and provide code for constructing them, as well as for writing code for simple simulation activities. We also outline the rationale for using profile likelihood-based standard errors in place of traditional model-based standard errors.

Chapter 3 addresses basic ordinary least squares (OLS) regression. Code structures are developed that will be used throughout the book. Least-squares regression is compared to full maximum likelihood methodology. Also discussed are problems related to missing data, reporting standard errors and confidence intervals, and to understanding S3 class modelling.

Chapter 4 relates to the theory and logic of programming generalized linear models (GLM). We spend considerable time analyzing the iteratively reweighted least squares (IRLS) algorithm, which has traditionally been used for the estimation of GLMs. We first demonstrate how to code specific GLM models as stand-alone functions, then show how all of them can be incorporated within a GLM covering algorithm. The object is to demonstrate the development of modular programming. A near complete GLM function, called `irls`, is coded and explained, with code for the three major Bernoulli and binomial models, Poisson, negative binomial, gamma, and inverse Gaussian models included. We also provide the function with a wide variety of post estimation statistics and a separate summary function for displaying regression results. Topics such as over-dispersion, offsets, goodness-of-fit, and residual analysis are also examined.

Chapter 5 develops traditional maximum likelihood methodology showing how to write modular code for one- and two-parameter GLMs as full maximum likelihood models. One parameter GLMs (function `ml_glm`) include binomial, Poisson, gamma, and inverse Gaussian families. Our `ml_glm2` function allows modelling of two-parameter Gaussian, gamma, and negative binomial regression models. We also develop a model that was not previously available in R, the heterogeneous negative binomial model, or NB-H. The NB-H model allows parameterization of the scale parameter as well as for standard predictor parameters.

Chapter 6 provides the logic and code for using maximum likelihood for the estimation of basic fixed effects and random effects models. We provide code for a conditional fixed effects negative binomial as well as a Gaussian random intercept model. We also provide an examination of the logic and annotated code for a working EM algorithm.

In the final chapter, Chapter 7, we address simulation as a method for estimating the parameters of regression procedures. We demonstrate how to construct synthetic models, then Monte Carlo simulation and finally how to employ Markov Chain Monte Carlo simulation for the estimation of Poisson regression coefficients, standard errors and associated statistics. In doing so we provide the basis of Bayesian modelling. We do not, however, wish to discuss Bayesian methodology in detail, but only insofar as the basic method can be used in estimating model parameters. Fully working annotated code is provided for the estimation of a Bayesian model with non-informative priors. The code can easily be adapted for the use of other data and models, as well as for the incorporation of priors.

Exercise questions are provided at the end of each chapter. We encourage the readers to try answering them. We have designed them so that they are answerable given the information provided in the chapter.

Our goal throughout has been to produce a clear and fully understandable volume on writing code for the estimation of statistical models using the foremost estimation techniques that have been developed by statisticians over the last half century. We attempt to use examples that will be of interest to researchers across disciplines.

As a general rule, we will include R code that the reader should be able to run, conditional on the successful execution of earlier code, and signal that code with the usual R prompts '<' and '+'. We will also include pseudo-code that provides some greater generality but should not be run as is. We omit the prompts for the pseudo-code to distinguish it from the executable code.

Acknowledgments

We wish to thank Rob Calver, statistics editor at Chapman & Hall/CRC (Taylor and Francis), for believing in this project, and for allowing us to write the book as we saw fit. Others whom we wish to thank include, [Hilbe] Alain Zuur, Highland Statistics, James Hardin (University of South Carolina), Robert Muenchen (University of Tennessee), and [Robinson] Mark Burgman (University of Melbourne), Jeff Gove (USDA FS), John Maindonald (ANU), Gordon Smyth (WEHI), and Murray Aitkin (University of Melbourne). We thank the R and LaTeX communities, and the authors and maintainers of Sweave, for these phenomenal resources.

Authoring books such as this one takes a great deal of writing and research time. However, most of our time was taken up in coding, testing, error checking, running models, re-coding, and so forth. This effort takes considerable time and patience, time that would otherwise be spent with our families. We therefore thank our families for not complaining about the times we were physically, as well as mentally, absent while working on this volume. Specifically, JMH wishes to acknowledge the support of his wife Cheryl, daughter Heather, sons Michael and Mitchell, grandsons Austin and Shawn, and Sirr, a white Maltese who keeps him company throughout the day when working on the computer. APR is grateful for the support of Grace, his son Felix, and Henry, a black-and-tan mutt who got walked to the beach far less often than he would like.

Joseph M. Hilbe
Florence, AZ, USA (hilbe@asu.edu)

Andrew P. Robinson
Melbourne, Australia (apro@unimelb.edu.au)

1

Programming and R

1.1 Introduction

The goal of this chapter is to introduce the reader to the programming tools that will be used in subsequent chapters. It therefore provides a highly selective review of R programming.

Users who have some exposure to data analysis and statistical packages that provide graphical user interfaces may be wary about such a seemingly bare-bones introduction to R. Many other data analysis products provide apparently straightforward importation of data, accompanied by attractive graphics and automated model fitting. Why is it useful to dig about in the tissue of the language? The reason is that, in our experience as statisticians, the provision to the analyst of data that are clean and ready to analyze is the exception rather than the rule. Invariably some pre-analysis processing is required. R provides a very flexible and powerful set of tools for the manipulation of data. Careful use of these tools will both ease the process and improve the transparency of preparing the data for suitable analysis. Therefore, close examination of the data manipulation facilities of R will benefit the analyst.

The definitive reference to R is the R Language Definition, which is freely available in PDF and HTML format on the R website, as well as being provided by default with each R installation. This work is continually updated by the volunteers that support R. It can be accessed via the `The R Language Definition` link on the front page of the html help file that is opened by the `help.start` function.

1.2 R Specifics

Making a definitive description of R is a tricky proposal, because R is multi-faceted and evolving. Therefore, we will tackle a simpler problem and describe R just as we will be treating it in this book. R, for the purposes of this book, is an interpreted, impure object-oriented programming language that provides many structures and functions that ease the importation, handling, and analysis of data, as well as reporting the outcome. Also for the purposes

1

of this book, R is software that the user interacts with via a command-line interface. The label R is commonly used to describe both the language and the software application that interprets it. R is more fully documented in readily available resources (e.g., R Development Core Team, 2012).

According to the R FAQ, the design of R has been heavily influenced by two other languages: it is very similar in appearance to Becker, Chambers & Wilks' S, and its underlying implementation and semantics are derived from Sussman's Scheme (Hornik, 2010). R is not uncommonly described as being "not unalike S."

R is an interpreted, as opposed to a compiled, language. An interpreted language is one for which the most common implementation involves translating the code to machine-executable commands and executing those commands one at a time. Re-running the program requires re-translation. A compiled implementation is one in which programs are written as collections of instructions, then converted to binary objects. Such binary objects then can be run many times without re-translation. We note in passing that R can run programs that have been compiled from other languages, such as C and FORTRAN. Furthermore, as of R version 2.13.0, a byte compiler is available, which provides useful although occasionally modest decreases in execution time.

As far as the user is concerned, the disadvantage of R being an interpreted language rather than a compiled language is that it is slower in execution than it would be if it were compiled. However, in the experience of the authors, the execution time of R is very rarely a bottleneck in analytical exercises.

The R software provides an interpreter, or listener. The listener accepts user input in the form of R code, then the software interprets the input, executes the instructions, and returns the output. In practice, the user types commands at the prompt, and R executes those commands, like this:

```
> 1 + 2
```

```
[1] 3
```

R, being somewhat like S, is somewhat object oriented (OO). Describing R's OO nature is complicated because of several factors: first, by the fact that there are different kinds of object orientation, and second, that R itself provides several implementations of OO programming. Indeed it is possible, although may be inefficient, for the user to ignore R's OO nature entirely. Hence R is an impure object-oriented language (according to the definition supplied by Craig, 2007, for example).

In this book we constrain ourselves to describing the implementation of S3 classes, which were introduced to S in version 3 (Chambers, 1992a). At the time of writing, R also provides S4 classes (Chambers, 1998), and the user-contributed packages *R.oo* (Bengtsson, 2003) and *proto* (Kates and Petzoldt, 2007). S3's object orientation is class-based, as opposed to prototype based. We will cover object-oriented programming (OOP) using R in more detail in Section 1.3.3. In the meantime, we need to understand that R allows the user to create and manipulate objects, and that this creation and manipulation is central to interacting efficiently with R.

1.2.1 Objects

Everything in R is an *object*. Objects are created by the evaluation of R *statements*. If we want to save an object for later manipulation, which we most commonly do, then we choose an appropriate name for the object and assign the name using the left arrow symbol `<-`. It is also possible to use the equals sign `=`; however, in this book we prefer `<-`. So object creation is, broadly, as follows.

```
name <- R statements
```

Valid object names may contain letters, digits and the two characters . and _, and must start with a letter or . (Chambers, 2008).

1.2.1.1 Vectors

We start with a vector object, of which there are six types: real, string, logical, integer, complex, and raw. Here we will focus on the first three types. We create a vector of three real numeric objects, which we shall call `wavelengths`, as follows:

```
> wavelengths <- c(325.3, 375.6, 411.1)
```

This code used the c function to concatenate the three numbers into a vector object, and then assigned the vector object a name: `wavelengths`. This vector object is a container for the three real numbers. We can print the object by just entering its name at the prompt.

```
> wavelengths
```

```
[1] 325.3 375.6 411.1
```

Every object in R has a *class*, which controls how the object can be manipulated. The class of the object can be determined (and set) using the `class` function.

```
> class(wavelengths)
```

```
[1] "numeric"
```

Classes are baked into base R, so knowing what they do and what they are for helps the user understand what R is doing. We will cover classes in greater detail later in this chapter. For the moment, we comment that knowing the class is very helpful.

We can create a vector of character strings in the same way:

```
> sentence <- c("This", "is", "a", "character", "vector")
> class(sentence)
```

```
[1] "character"
```

We remark that R has some wonderful character-handling functions such as `paste`, `nchar`, `substr`, `grep`, and `gsub`, but their coverage is beyond the scope of this book.

Many operations are programmed so that if the operation is called on the vector, then it is efficiently carried out on each of its elements. For example,

```
> wavelengths / 1000
```

```
[1] 0.3253 0.3756 0.4111
```

This is not true for all operations; some necessarily operate on the entire vector, for example, `mean`,

```
> mean(wavelengths)
```

```
[1] 370.6667
```

and `length`.

```
> length(wavelengths)
```

```
[1] 3
```

R also provides a special class of integer-like objects called a *factor*. A factor is used to represent a categorical (actually nominal, more precisely) variable, and that is the reason that it is important for this book. The object is displayed using a set of character strings, but is stored as an integer, with a set of character strings that are attached to the integers. Factors differ from character strings in that they are constrained in terms of the values that they can take.

```
> a_factor <- factor(c("A","A","B","B","B","C"))
> a_factor
```

```
[1] A A B B B C
Levels: A B C
```

Note that when we printed the object, we were told what the *levels* of the factor were. These are the only values that the object can take. Here, we try to turn the 6th element into a Z.

```
> a_factor[6] <- "Z"
> a_factor
```

```
[1] A    A    B    B    B    <NA>
Levels: A B C
```

We failed: R has made the 6th element *missing* (NA) instead. Note that we accessed the individual element using square brackets. We will expand on this topic in Section 1.2.1.2.

The levels, that is the permissible values, of a factor object can be displayed and manipulated using the `levels` function. For example,

```
> levels(a_factor)

[1] "A" "B" "C"
```

Now we will turn the second element of the levels into Bee.

```
> levels(a_factor)[2] <- "Bee"
```

The consequence of this operation is that when we now print the factor, the levels have been changed.

```
> a_factor

[1] A    A    Bee  Bee  Bee  <NA>
Levels: A Bee C
```

One challenge that new R users often have with factors is that the functions that are used to read data into R will make assumptions about whether the intended class of input is factor, character string, or integer. These assumptions are documented in the appropriate help files, but are not necessarily obvious otherwise. This is especially tricky for some data where numbers are mixed with text, sometimes accidentally. Manipulating factors as though they really were numeric is often perilous. For example, they cannot be added.

```
> factor(1) + factor(2)

[1] NA
```

The final object class that we will describe is the logical class, which can be thought of as a special kind of factor with only two levels: TRUE and FALSE. Logical objects differ from factors in that mathematical operators can be used, e.g.,

```
> TRUE + TRUE

[1] 2
```

It is standard that TRUE evaluates to 1 when numerical operations are applied, and FALSE evaluates to 0. Logical objects are created, among other ways, by evaluating logical statements, for example

```
> 1:4 < 3
```

```
[1]   TRUE   TRUE FALSE FALSE
```

Logical objects can be manipulated by the usual logical operators, *and* &, *or* |, and *not* !. Here we evaluate TRUE (or) TRUE.

```
> TRUE | TRUE
```

```
[1] TRUE
```

In evaluating this statement, R will evaluate each logical object and then the logical operator. An alternative is to use && or ||, for which R will evaluate the first expression and then only evaluate the second if it is needed. This approach can be faster and more stable under some circumstances. However, the latter versions are not vectorized.

The last kind, but not class, of object that we want to touch upon is the missing value, NA. This is a placeholder that R uses to represent known unknown data. Such data cannot be lightly ignored, and in fact R will often retain missing values throughout operations to emphasize that the outcome of the operation depends on the missing value(s). For example,

```
> missing.bits <- c(1, NA, 2)
> mean(missing.bits)
```

```
[1] NA
```

If we wish to compute the mean of just the non-missing elements, then we need to provide an argument to that effect, as follows (see Section 1.2.3.1, below).

```
> mean(missing.bits, na.rm = TRUE)
```

```
[1] 1.5
```

Note that the missing element is counted in the length, even though it is missing.

```
> length(missing.bits)
```

```
[1] 3
```

We can assess and set the missing status using the is.NA function.

1.2.1.2 Subsetting

In the previous section we extracted elements from vectors. Subsets can easily be extracted from many types of objects, using the square brackets operator or the subset function. Here we demonstrate only the former. The square brackets take, as an argument, an expression that can be evaluated to either an integer object or a logical object. For example, using an integer, the second item in our sentence is

```
> sentence[2]

[1] "is"
```

and the first three are

```
> sentence[1:3]

[1] "This" "is"   "a"
```

Note that R has interpreted 1:3 as the sequence of integers starting at 1 and concluding at 3. We can also exclude elements using the negative sign:

```
> sentence[-4]

[1] "This"   "is"      "a"       "vector"

> sentence[-(2:4)]

[1] "This"   "vector"
```

An example of the use of a logical expression for subsetting is

```
> wavelengths > 400

[1] FALSE FALSE   TRUE

> wavelengths[wavelengths > 400]

[1] 411.1
```

These subsetting operations can be nested. Alternatively, the intermediate results can be stored as named objects for subsequent manipulation. The choice between the two approaches comes down to readability against efficiency; creating interim variables slows the execution of the code but allows commands to be split into more easily readable code chunks.

1.2.2 Container Objects

Other object classes are containers for objects. We cover two particularly useful container classes in the section: lists and dataframes.

1.2.2.1   Lists

A list is an object that can contain other objects of arbitrary and varying classes. A list is created as follows.

```
> my.list <- list(number = 1, text = "alphanumeric")
```

Elements can be manipulated or extracted from the list using the double square bracket symbol, or the $ symbol, as follows.

```
> my.list[[1]]
```

```
[1] 1
```

```
> my.list$text
```

```
[1] "alphanumeric"
```

```
> my.list[[1]] <- 2
> my.list
```

```
$number
[1] 2
```

```
$text
[1] "alphanumeric"
```

Empty lists can be conveniently created using the **vector** function, as follows.

```
> capacious.empty.list <- vector(1000, mode = "list")
```

This list can then be used as a container for the outcome of a loop. Functions like `lapply` and `sapply` allow for elegant, element-wise evaluation of functions upon lists. For example, to determine the class of each element in a list, we can use the following code:

```
> sapply(my.list, class)
```

```
     number        text
  "numeric" "character"
```

Lists are very useful devices when programming functions, because functions in R are only allowed to return one object. Hence, if we want a function to return more than one object, then we have it return a list that contains all the needed objects.

1.2.2.2 Dataframes

Dataframes are special kinds of lists that are designed for storing and manipulating datasets as they are commonly found in statistical analysis. Dataframes typically comprise a number of vector objects that have the same length; these objects correspond conceptually to columns in a spreadsheet. Importantly, the objects need not be of the same class. Under this setup, the i-th unit in each column can be thought of as belonging to the i-th observation in the dataset.

There are numerous ways to construct dataframes within an R session. We find the most convenient way to be the `data.frame` function, which takes objects of equal length as arguments and constructs a dataframe from them. If the arguments are named, then the names are used for the corresponding variables. For example,

```
> example <- data.frame(var.a = 1:3,
+                       var.b = c("a","b","c"))
> str(example)

'data.frame':        3 obs. of  2 variables:
 $ var.a: int  1 2 3
 $ var.b: Factor w/ 3 levels "a","b","c": 1 2 3
```

If the arguments are of different lengths, then R will repeat the shorter ones to match the dimension of the longest ones, and report a warning if the shorter are not a factor (in the mathematical sense) of the longest.

```
> data.frame(var.a = 1,
+            var.b = c("a","b","c"))

  var.a var.b
1   1     a
2   1     b
3   1     c
```

Furthermore, as is more commonly used in our experience, when data are read into the R session using one of the `read.xxx` family of functions, the created object is a dataframe.

There are also numerous ways to extract information from a dataframe, of which we will present only two: the square bracket [operator and `subset`.

The square bracket operator works similarly as presented in Section 1.2.1.2, except instead of one index it now requires two: the first for the rows, and the second for the columns.

```
> example[2, 1:2]

  var.a var.b
2   2     b
```

A blank is taken to mean that all should be included.

```
> example[2,]

  var.a var.b
2   2    b
```

Note that this operation is calling a function that takes an object and indices as its arguments. As before, the arguments can be positive or negative integers, or a logical object that identifies the rows to be included by TRUE.

Extracting data using the subset function proceeds similarly, with one exception: the index must be logical; it cannot be integer.

```
> subset(example, subset = var.a > 1, select = "var.b")

  var.b
2   b
3   c
```

It is worth noting that storing data in a dataframe is less efficient than using a matrix, and manipulating dataframes is, in general, slower than manipulating matrices. However, matrices may only contain data that are all the same class. In cases where data requirements are extreme, it may be worth trying to use matrices instead of dataframes.

1.2.3   Functions

We write functions to enable the convenient evaluation of sets of expressions. Functions serve many purposes, including, but not limited to, improving the readability of code, permitted the convenient re-use of code, and simplifying the process of handling intermediate objects.

In R, functions are objects, and can be manipulated as objects. A function comprises three elements, namely: a list of arguments, a body, and a reference to an environment. We now briefly describe each of these using an example. This trivial function sums its arguments, and if the second argument is omitted, it is set to 1.

```
> example.ok <- function (a, b = 1) {
+    return(a + b)
+ }
> example.ok(2,2)

[1] 4

> example.ok(2)

[1] 3
```

We can now examine the pieces of the function by calling the following functions. The formals are the arguments,

```
> formals(example.ok)
```

$a

$b
[1] 1

the body is the R code to be executed,

```
> body(example.ok)
```

```
{
    return(a + b)
}
```

and the environment is the parent environment.

```
> environment(example.ok)
```

```
<environment: R_GlobalEnv>
```

We describe each of these elements in greater detail below.

1.2.3.1 Arguments

The arguments of a function are a list (actually a special kind of list, called a *pairlist*, which we do not describe further) of names and, optionally, expressions that can be used as default values. So, the arguments might be for example x, or x = 1, providing the default value 1, or indeed x = *some expression*, which will be evaluated if needed as the default expression.

The function's arguments must be valid object names, that is, they may contain letters, digits and the two characters . and _, and must start with either a letter or the period . (Chambers, 2008). If an expression is provided as part of the function definition, then that expression is evaluated and used as the default value, which is used if the argument is not named in the function call.

```
> example <- function(a = 1) a
> example()
```

```
[1] 1
```

Note that the argument expressions are not evaluated until they are needed — this is lazy evaluation. We can demonstrate this by passing an expression that will result in a warning when evaluated.

```
> example <- function(a, b) a
> example(a = 1)
```

```
[1] 1
```

```
> example(a = 1, b = log(-1))
```

```
[1] 1
```

The absence of warning shows that the expression has not been evaluated.

It is important to differentiate between writing and calling the function when thinking about arguments. When we write a function, any arguments that we need to use in that function must be named in the argument list. If we omit them, then R will look outside the function to find them. More details are provided in Section 1.2.3.3.

```
> example <- function(a) a + b
```

```
> example(a = 1)
```

```
Error in example(a = 1) : object 'b' not found
```

Now if we define b in the environment in which the function was created, the parent environment, then the code runs.

```
> b <- 1
> example(a = 1)
```

```
[1] 2
```

Note that R searched the parent environment for b.

An exception is that if we want to pass optional arbitrary arguments to a function that is called within our function, then we use the ... argument. Below, note how **example** requires arguments a and b, but when we call it within our **contains** function we need to provide only a and the dots.

```
> example <- function(a, b) a + b
> contains <- function(a, ...) example(a, ...)
> contains(a = 1, b = 2)
```

```
[1] 3
```

Our **contains** function was able to handle the argument by passing it to **example**, even without advance warning.

In calling a function, R will match the arguments by name, position, or both name and position. For example,

```
> example(1, 2)
```

```
[1] 3

> example(1, b = 2)

[1] 3
```

In general we find it safest to name all the arguments that are included in the function call. Note that R will allow partial matching of arguments, but we have not found it useful so we do not cover it further here.

1.2.3.2 Body

The body of the function is a collection of R expressions that are to be executed. If there is more than one expression, then the expressions are contained within curly braces. For single expression functions the inclusion of the braces is a matter of taste.

The most common use of functions is to return some kind of object. Rarely, functions will do other things than return an object; these are called *side effects*, and should be used thoughtfully. An example of a useful side effect is opening a graphics device and plotting some data. An example of a risky side effect is the alteration of global variables from within the function. Here we create an object and then alter it from within a function.

```
> a <- 1
> a

[1] 1

> risky <- function(x) a <<- x
> risky(2)
> a

[1] 2
```

If the **return** function is omitted, then the last value that is computed in the function is returned.

```
> example.ok <- function (a, b = 1) {
+    a + b
+ }
> example.ok(2)

[1] 3
```

The returned object can be any R object, and can be manipulated *in situ* as a matter of programming taste. Note that we write of the returned object in the singular. If the function is to create more than one object, then the returned objects must be collected in a list, that is, the function must return a single object that itself can contain multiple objects. An example follows, in which we explicitly name the elements of the list to ease later extraction.

```
> example.ok <- function (a, b = 1) {
+   return(list(a = a, b = b, sum = a + b))
+ }
> example.ok(2)

$a
[1] 2

$b
[1] 1

$sum
[1] 3
```

Note that the returned object is a list, so the objects that it contains can be extracted using the usual list protocol.

```
> example.ok(2)$sum

[1] 3

> example.ok(2)[[3]]

[1] 3
```

1.2.3.3   Environments and Scope

The enclosing environment is the environment in which the function is *created*, not the environment in which the function is *executed*. We need to know about the environment because it tells us what objects are available to the function when the function is executed.

R is lexically scoped. This means that the objects that are available to a function at the time of its execution, in addition to its arguments and the variables created by the function itself, are those that are in the environment in which the function was created, the environment's enclosing environment, and so on up to the global environment. A brief example follows.

```
> x <- 100
> scope.fn <- function() x / 10
> scope.fn()

[1] 10
```

We see that even though x is declared outside the function, and is not provided to the function via the argument list, it is still available.

We now provide a more detailed example of how scoping works in R, in the context of this book. Here we have a function, **variance.binomial**, that will be called by another function, **irls**. At the time of writing

variance.binomial we know that it will need to be one among a number
of variance functions, each of which will have different arguments (the specific
details of the implementation will become obvious after study of Section 1.3.4).

For example, the Poisson distribution needs only the mean mu whereas the
binomial distribution needs mu and the count m. We would like to try to rely
on scoping to get the m parameter when it is needed and ignore it when it is
not. We start with a variance function that includes mu as an argument.

```
> variance.binomial <- function(mu) mu * (1 - mu / m)
```

Then we write the calling function, here greatly simplified(!), and again ignoring the specific S3 details.

```
> irls <- function(mu) {
+   m <- 100
+   variance.binomial(mu)
+ }
```

But when we try to run this function, m is not available within
variance.binomial, even though it is available within irls. This is because
variance.binomial was created in the global environment, not in the environment created by irls.

```
> irls(0.5)
```

```
Error in variance.binomial(mu, m) : object 'm' not found
```

We can verify this observation as follows:

```
> environment(variance.binomial)
```

```
<environment: R_GlobalEnv>
```

We can solve this problem in two ways. First, we can ensure that all possible arguments of interest are identified in the definition of the function that
is to be called internally, for example as follows.

```
> irls <- function(mu) {
+   m <- 100
+   variance.binomial(mu, m)
+ }
> variance.binomial <- function(mu, m) mu * (1 - mu / m)
> irls(0.5)
```

```
[1] 0.4975
```

Second, we can move the definition of the function that is to be called to be
within the calling function, as follows.

```
> irls <- function(mu) {
+    variance.binomial <- function(mu) mu * (1 - mu / m)
+    m <- 100
+    variance.binomial(mu)
+    }
> irls(0.5)
```

[1] 0.4975

Each solution has its drawbacks. In the first case, we have to include all the arguments that we might want. As we shall see in Section 4.7, this is a bit messy. For example, we have to pass arguments to the function that are simply passed to communicate the class to the generic, or we have to explicitly copy the class information to the object that is being passed via the arguments. Neither solution is elegant. In the second case we have to rewrite the function in order to expand the functionality of the code, which breaks the modularity. We elected the first, messy, modular solution.

1.2.4   Matrices

A matrix is a rectangular array of items, formatted in terms of rows (i) and columns (j). A matrix is defined by the values of i and j, which are called the dimensions of the matrix. Therefore, a matrix with 3 rows and 4 columns is a (3,4), or 3×4, dimension matrix. When the dimensions of a matrix are identical, for example 2×2, it is known as a square matrix.

In R, a *matrix* is just a vector with a special attribute called dim, which contains the size of the matrix in *row–column* format. As we shall see, R changes the way that it carries out arithmetical operations for matrices compared with vectors. Because a matrix is stored as a single vector, its values must all be stored in the same mode. For example, the values must all be numeric, all character, or all logical.

There are numerous ways to create matrices, of which we show only a few. We begin by creating and printing a 2×2 matrix.

```
> mat1 <- matrix(1:4, nrow=2)
> mat1
```

```
     [,1] [,2]
[1,]    1    3
[2,]    2    4
```

We can verify that it has a dimension attribute by using the attributes function, or more simply using the dim function.

```
> attributes(mat1)
```

$dim
[1] 2 2

```
> dim(mat1)
```

```
[1] 2 2
```

We now mention *attributes* briefly, and direct the interested reader to Chapter 5 of Becker et al. (1988). Attributes provide a means by which useful information can be carried along with objects. For example, we may wish to retain the units of some light measurements without wishing to formally define a new class of objects. We could do so by setting an attribute. In general, the attributes of an object can be changed or read using the `attr` function, as follows.

```
> attr(wavelengths, "units") <- "micrometres"
> attributes(wavelengths)
```

```
$units
[1] "micrometres"
```

As matrices are just a vector with an attribute, any functions that work with vectors will work identically with matrices. For example, we can sum the elements of a matrix by using the `sum` function.

```
> sum(mat1)
```

```
[1] 10
```

Furthermore, there are special matrix-handling functions. For example, if we wish to sum the columns (or rows), then we use the efficient `colSums` (or `rowSums`) function.

```
> colSums(mat1)
```

```
[1] 3 7
```

The standard arithmetic functions operate on the matrices as though they were vectors — that is, element by element. We show some of these functions below, along with the `diag` function, which provides another way to make a useful matrix.

```
> (mat2 <- diag(2))
```

```
     [,1] [,2]
[1,]    1    0
[2,]    0    1
```

```
> mat1 + mat2
```

```
     [,1] [,2]
[1,]    2    3
[2,]    2    5
```

```
> mat1 * mat2
```

```
     [,1] [,2]
[1,]   1    0
[2,]   0    4
```

R also provides matrix-specific functions that cover such operations as computing the product of two matrices. The matrix product operator is %*%. Here we take the matrix product of mat1 with mat2.

```
> mat1 %*% mat2
```

```
     [,1] [,2]
[1,]   1    3
[2,]   2    4
```

It is worth noting that the crossprod function serves the same general purpose as %*%, and is often faster, so we will prefer to use it and its companion tcrossprod in the rest of the book, except where programming clarity is reduced.

```
> tcrossprod(mat1, mat2)
```

```
     [,1] [,2]
[1,]   1    3
[2,]   2    4
```

Matrix inversion is available using the solve function, as follows.

```
> solve(mat1)
```

```
     [,1] [,2]
[1,]   -2  1.5
[2,]    1 -0.5
```

We can check that the matrix is indeed the inverse by

```
> mat1 %*% solve(mat1)
```

```
     [,1] [,2]
[1,]   1    0
[2,]   0    1
```

The Moore–Penrose generalized inverse is available from the ginv function in the *MASS* package.

R also provides built-in functions to perform back-solving (backsolve), singular-value decomposition (svd), Choleski decomposition (col), and QR-decomposition (qr), some of which we shall use in Chapter 3.

1.2.5   Probability Families

R provides a suite of functions that enable the direct use of many popular probability families. These functions are extremely useful for statistical inference, for example when estimating parameters using maximum likelihood and when providing interval estimates for parameters.

The template for any given probability family named *family* is a set of four functions: the probability distribution function, called dfamily; the cumulative mass (or density) function (CDF), called pfamily; the inverse CDF, called qfamily; and a pseudo-random number generator, called rfamily. So, the four functions in the case of the Poisson distribution are: dpois, ppois, qpois, and rpois.

In each case, the functions require arguments that represent the parameters of the distribution, as well as other arguments that are specific to the function's intention. For example, the dfamily and pfamily functions need the value at which the PDF and CDF should be calculated, and rfamily function requires the number of observations to be generated. Some families have default values for the arguments that represent the parameters. For example, the default arguments that represent the parameters of the distribution for the normal family are mean = 0 and sd = 1. In contrast, the binomial family has no default values.

Most of the in-built probability functions call highly efficient pre-compiled C code, so the calculations proceed very quickly. The help files provide citations to the algorithms that are used, and of course, the source code can also be examined to see how those algorithms are implemented.

Here we provide an example of use of the functions for the normal distribution family. To calculate the probability density of a random standard normal variate at 0, where the standard normal distribution has zero mean and unit variance, use

```
> dnorm(0, mean = 0, sd = 1)
```

```
[1] 0.3989423
```

The following is equivalent.

```
> dnorm(0)
```

```
[1] 0.3989423
```

The PDF function has an additional argument: log = FALSE. This argument flags whether to return the probability or the log of the probability, and is very useful when using the PDF function for obtaining maximum-likelihood estimates, for example. Typically the code that is used to return the log of the probability is written especially for that purpose, which improves speed and reduces rounding error. Hence, calling dfamily(x, ..., log = TRUE) is much preferable to calling log(dfamily(x, ...)).

To calculate the probability that a random standard normal variate will be less than 0.5, use

```
> pnorm(0.5)
```

```
[1] 0.6914625
```

To find the value that is the upper 0.975 quantile of the normal distribution with mean 2 and variance 4, we use

```
> qnorm(0.975, mean = 2, sd = 2)
```

```
[1] 5.919928
```

Finally, to generate 5 pseudo-random normal variates from the same distribution, we use

```
> rnorm(5, mean = 2, sd = 2)
```

```
[1] 2.8737530 2.2841174 0.4283859 1.6636226 1.7908804
```

Although R provides a wealth of in-built, efficient functions for manipulating probability families, it is possible that an analyst may need to write custom functions for new families. Depending on the application, the functions may either be relatively brief, without much by the way of error-checking infrastructure, or they could be quite complicated, with code that is optimized for speed, stability, and numerical accuracy, and detailed checks for input legality, etc.

Here we construct simple probability functions for the following PDF, which we shall call Watson's function.

$$f(x; \theta) = \frac{1 + \theta}{\theta \left(1 + \frac{x}{\theta}\right)^2} \quad 0 < x \leq 1; \quad \theta > 0 \tag{1.1}$$

```
> dwatson <- function(x, theta) {
+    (1 + theta) / theta / (1 + x / theta)^2
+ }
```

There is a closed-form expression for the CDF;

$$F(x) = (1 + \theta) \times \left(1 - \left(1 + \frac{x}{\theta}\right)^{-1}\right), \tag{1.2}$$

but we prefer to use a more general solution for our simple example.

N.B.: An early draft of this chapter included an example PDF that was defined on the entire real line; however, constructing stable, simple, and general user functions proved to be trickier for that example than we had hoped. Closed-form expressions are certainly preferred if they exist, because they are easier to program in numerically stable ways, and also may well be vectoriz-able. The exercise was also a useful lesson in just how much numerical tech-nology has gone into the in-built functions that we tend to take for granted.

Here, we use the R function `integrate` to numerically integrate the func-tion that represents the PDF, in order to create a CDF.

```
> pwatson <- function(q, theta) {
+    integrate(function(x) dwatson(x, theta),
+              lower = 0, upper = q)$value
+  }
```

As for the CDF, it is preferable to express the inverse CDF in closed form exactly, or even to a known approximation. If that is impossible, then we can use the `uniroot` function to locate the value of x at which the CDF $F_X(x)$ is equal to the desired value. Our function cannot be vectorized, so is not efficient to use, but it is sufficient for our purposes.

```
> qwatson <- function(p, theta) {
+    uniroot(function(x) pwatson(x, theta) - p,
+            lower = .Machine$double.eps, upper = 1)$root
+ }
```

The use of `.Machine$double.eps` in an argument should be explained. R has a number of computer-specific performance-related constants that are accessible within the `.Machine` list object. Here we are asking R to set the lower limit to be the 'smallest positive floating-point number x such that $1 + x \neq 1$' (see `?.Machine`).

It is sensible to apply some checks to the function to be sure that it is a PDF and that the transcription to R code is correct. In order to be usable as a PDF, a function must be non-negative across the support of the random variable, and must integrate to one.

Integrability to one should be checked for a number of values of the pa-rameter; here we try only one of them.

```
> pwatson(1, theta = 1)
```

```
[1] 1
```

Finally, we can construct a pseudo-random number generator by provid-ing a pseudo-random uniform (0,1) variate as the quantile argument to the `qwatson` function.

Recall that `qwatson` was not vectorized. In that case, we have to explicitly call it as many times as we wish pseudo-random numbers. Here we do so using the convenient `sapply` function.

```
> rwatson <- function(n, theta) {
+    sapply(runif(n), function(x) qwatson(x, theta))
+ }
```

We call the function as follows.

```
> rwatson(5, 1)
```

```
[1] 0.04027990 0.31159946 0.08243589 0.14475373 0.22647895
```

The resulting generator is slow compared with the in-built functions to generate random numbers, e.g., below we compare the execution time of the R function that generates normal variates with the execution time of our function that generates Watson variates, to the detriment of the latter.

```
> system.time(rnorm(1000))

   user  system elapsed
      0       0       0

> system.time(rwatson(1000,1))

   user  system elapsed
  0.329   0.004   0.333
```

Despite the time difference, generating 1,000 variates took less than a second on a modest laptop. As a final exercise, we coded the closed-form version of the inverse-CDF to see what the time difference would be compared with our earlier inefficient solution.

```
> qwatson1 <- function(p, theta) {
+    (1 / ( 1 - p / (1 + theta) ) - 1) * theta
+ }
> rwatson1 <- function(n, theta) {
+    qwatson1(runif(n), theta)
+ }
> system.time(rwatson1(1000,1))

   user  system elapsed
  0.001   0.000   0.000
```

This output shows the predictable result that using vectorized arithmetic operations on closed-form expressions is considerably faster than applying a root-finding function to a function that uses numerical integration.

1.2.6   Flow Control

So far, the code that we have constructed will be evaluated in a linear, sequential fashion — one expression after another. Often we want to exercise further control over the execution of different bits of code. This might be wanting to choose between expressions conditional on the outcome of a logical test, to execute a collection of expressions a given number of times, or with minor modifications, or to execute a collection of expressions until a condition changes. R provides powerful functions that allow each of these behaviors.

1.2.6.1   Conditional Execution

We can direct R to execute an expression conditional on the outcome of a logical test using the `if` function, or to choose between two expressions using the `if` and `else` functions.

The `if` function works alone in the following general pattern (this particular code will not run):

```
if (test) { body }
```

where `test` is an expression that evaluates to a logical object of length 1, and `body` is one or more R expressions, wrapped in braces. If `body` comprises only one R expression, then the braces may be omitted.

The `if` / `else` combination works as follows (this particular code will not run):

```
if (test) {
    true.body
    } else {
    false.body
}
```

Note that the call to `else` is directly after the brace that closes the expressions to be executed if `test` is true. This is important because R will not know otherwise that an `else` is included. So, the following snippet will emphatically fail because of the locations of the parentheses.

```
if (test) { true.body }
else { false.body }
```

We can see that upon the closing of the brace after `true.body`, R has received a complete expression.

For vectorized conditional execution, we use the `ifelse` function.

1.2.6.2   Loops

Sometimes we wish to execute a collection of statements a set number of times, either as they are, or varying a portion of them in predictable ways. We use loops for this purpose. R provides two looping functions: `for` and `while`.

We use `for` to repeat a collection of expressions a given number of times. It works as follows (this particular code will not run).

```
for (index in sequence) { body }
```

Here, `in` is the word in(!) and `index` identifies an object that will take the values in `sequence` during the evaluation of `body`. For example, the following code prints the integers from one to two.

```
> for (i in 1:2) cat(i, "\n")
```

```
1
2
```

It is worth noting that **sequence** can be any expression that evaluates to a vector. This flexibility can lead to code that is easier to read.

```
> for (i in c("a","b")) cat(i, "\n")
```

```
a
b
```

When evaluation of the loop is complete, the value of **index** is retained as the last in **sequence**.

```
> i
```

```
[1] "b"
```

It is important to be sure that **sequence** evaluates correctly. The **for** function is equipped to handle **NULL** sequences without an error, but that condition cannot be delivered using the colon and integers. Therefore the use of **seq** is a more robust equivalent (see Chambers, 2008).

```
> seq(2)
```

```
[1] 1 2
```

```
> seq(NULL)
```

```
integer(0)
```

Sometimes the intention is to apply a collection of expressions to all of the objects in a list. If so, it is often tempting to store the output in a vector or list that is grown with each iteration. This temptation should be resisted. Instead, the storage should be created before the loop is evaluated. We use a list in the following example, although a vector would be slightly more efficient and is all that the code requires.

```
> output <- vector(length = 10, mode = "list")
> for (i in 1:10) output[[i]] <- paste("Element", i)
> output[[1]]
```

```
[1] "Element 1"
```

Alternatively, in many simple cases the **lapply** function and its siblings wrap the operations into a single, compact call.

```
> output <- lapply(1:10, function(i) paste("Element", i))
> output[[1]]
```

```
[1] "Element 1"
```

When the goal is to continually evaluate a set of expressions until a condition changes, we use the `while` function (this particular code will not run).

```
while (test) { body }
```

It is important to ensure that the expressions in `body` update the object(s) that are evaluated in `test`. An example of its usage that is sufficient for our purposes follows.

```
> i <- 1
> while (i < 2) {
+    cat(i, "\n")
+    i <- i + 1
+ }
```

```
1
```

1.2.7 Numerical Optimization

Numerical optimization rests at the foundation of statistical model estimation. Modelling algorithms are used to estimate parameters, which are expressed as coefficients or slopes. Regression slopes are the solutions to partial derivatives when set to zero. Therefore the process of estimation is simply one of maximization, or depending on the situation, minimization. Maximum likelihood estimation (MLE), the foremost method used by statisticians for the estimation of regression parameters, is a maximization procedure, just with a very particular objective function. In this section we briefly cover numerical optimization as it most commonly is applied to statistical model fitting.

Suppose we provide a function $f(x) = 2x^2 - \log(3x)$ and we wish to determine the value that maximizes or minimizes the function. We know from elementary calculus that we need only take the derivative of the function with respect to x, set the derivative to zero, and solve for the unknown parameter.

$$\frac{df}{dx} = 4x - \frac{3}{3x} \tag{1.3}$$

We set this equation to equal 0, so $4x^2 - 1 = 0$ and so $x = \sqrt{0.25} = 0.5$. In this case the function is univariate, so we can also plot the trajectory and determine the location of the value by inspection, using a graphic device such as is presented in Figure 1.1.

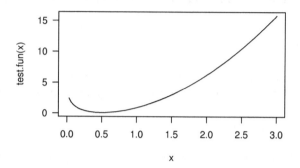

FIGURE 1.1
Example function.

```
> test.fun <- function(x) 2*x^2 - log(3*x)
> curve(test.fun, from = 0, to = 3)
```

The study of algorithms used for maximization or minimization is generally referred to in mathematics as numerical optimization. In this section we shall provide a brief look at how optimization relates to statistical model estimation as discussed in this text. The exception is estimation based on Markov Chain Monte Carlo (MCMC) methods, which underlies much of modern Bayesian methodology. We discuss MCMC methods and the Metropolis–Hastings sampling algorithm in the final chapter.

A variety of optimization algorithms have been used for maximization (or minimization). The most commonly known algorithms include golden-section, fixed-point, line search, steepest ascent, Nelder–Mead, Newton, and Newton–Raphson. Newton–Raphson is the most commonly used algorithm, and is the basis of the algorithm most commonly used for maximum likelihood estimation, so we focus upon it here. The interested reader can learn more from Nocedal and Wright (2006) and Jones et al. (2009).

The Newton–Raphson algorithm is the most adapted search procedure used for the estimation of statistical models. It is based on a manipulation of the two-term Taylor expansion. In this form, the Taylor expansion appears as

$$f(x) = f(x_0) + (x - x_0)f'(x_0) + \dots \qquad (1.4)$$

We want to find the value of x for which $f'(x) = 0$. So, we take the two-term Taylor expansion of $f'(x)$, instead of $f(x)$.

$$f'(x) = f'(x_0) + (x - x_0)f''(x_0) + \dots \qquad (1.5)$$

Now, setting $f'(x) = 0$ and rearranging, we get

$$x = x_0 - \frac{f'(x_0)}{f''(x_0)}, \qquad (1.6)$$

which provides us a way to convert an estimate of x, say x_n, into one that should be closer to a local optimum, say x_{n+1}, as follows

$$x_{n+1} = x_n - \frac{f'(x_n)}{f''(x_n)}, \qquad (1.7)$$

We continue to discuss optimization in a more specific setting, that of maximum likelihood, in Section 2.3.2.

R has a variety of optimization procedures, but one foremost function provides the key algorithms, namely optim, which we have already introduced. The definition for the optim function is

```
optim(par, fn, gr = NULL, ...,
  method = c("Nelder-Mead",BFGS","CG","L-BFGS-B","SANN","Brent"),
  lower = -Inf, upper = Inf,
  control = list(), hessian = FALSE)
```

The first argument to the function is par, which is a vector of the starting values for each of the free parameters. fn is the function to be minimized, and the default method is Nelder–Mead (Nelder and Mead, 1965). With hessian = TRUE designated, the Hessian is estimated, which allows estimation of standard errors. We continue this topic with a tighter focus on optimization for generalized linear models in Section 2.3.2.

1.3 Programming

1.3.1 Programming Style

Programming style is an intersection of individual taste, the characteristics of the language being used, the programmer's experience with the language (and other languages), and the problem at hand. Programming style shows itself not only in the structure of computer code, but also the decisions that the code forces the user to make, and the information that it provides.

For example, the lm function in R will quietly ignore missing values, without especial feedback to the user. Information about missingness is reported by the print.summary.lm function, which is called when a summary of the fitted model is printed. Alternatives would be for the lm function to provide a message, a warning, or an error. Each of these options, including the default, reflects a legitimate choice about an important decision, which is how to handle missing values in linear regression. In this book, we elect to place as much responsibility on the code user as we can. That means, for example, that our functions will throw errors when data are missing, so the user has to deal proactively with the problem of missing data. This choice is particularly driven by the shock of the second author upon discovering that, after years of blithely assuming that the tree heights were missing completely at random in the ufc dataset (included in the *msme* package), there is strong evidence that they are not (see Chapter 4).

Throughout the construction of the code for this book, we have found it most efficient for us to start with very simple prototypes that satisfy simple needs. We then add complications until the code is useful, at which point, we usually look for opportunities to unify or simplify the functions. Our guiding principle has been to simplify the problem until the answer is obvious, and then complicate it until the answer is useful.

The other element of style that we wish to discuss, which is another way of saying that we wish to defend, is the structure and number of functions that we have written in order to solve our problems. We have tried to develop the structure and number of functions that we think best represents the unity and elegance of the underlying statistical theory. That is, our interest is pedagogic. For example, in Chapter 5 we need to maximize the joint log-likelihood of a sample of data, conditional on a model, as a function of the model parameters. We elect to write separate functions for calculating the joint log-likelihood, then for summing it, and for maximizing it. Each function calls the others as necessary. This approach is not efficient use of computer memory or time. Nevertheless, the approach has two benefits: first, it allows us to map the objects to the problem as we think that it is most easily understood, and second, it does allow us to re-use the functions elsewhere without alteration. So, extending the code to enable fitting an entirely different model could be as simple as writing two one-line functions. Furthermore, we re-use the joint log-likelihood code for computing the deviance residuals of the fitted model. Doing so means that the deviance residuals can be computed for a new model merely by provision of the joint log-likelihood of the model, whereas otherwise the deviance residuals for the new model would need their own separately coded function. Finally, the pattern somewhat follows the admirable UNIX philosophy, which, broadly sketched, involves the development of simple functions that can be chained together to solve complicated problems.

Perhaps the tone of this section is unnecessarily defensive. We think that the core message should be that programming style is important, and the programmer should take time to reflect on the higher-level structure of code in the context of the immediate need and also future applications, ideally before too many bytes are committed.

1.3.2   Debugging

Debugging is the vernacular name given to the process of solving computer-related problems. Structured approaches to solving problems will be profitable for debugging exercises. Broadly speaking, we use the following steps when confronted by a problem.

1. Enumerate the conditions that identify the problem

2. Read widely

3. Experiment

   (a) Develop a hypothesis
   (b) Design an experiment to test the hypothesis
   (c) Carry out the experiment and record the results
   (d) Draw conclusions

4. Try to simplify the problem.

We particularly recommend (Polya, 1988) as useful reading.

The best tools for debugging in R depend on the nature of the problem. When R throws an error or an unexpected result, we find the `str` function *invaluable*. We could broadly estimate that more than two thirds of the debugging exercises that we have undertaken for our own work or on behalf of others have been resolved by examining the output of `str` and comparing it critically with the assumptions made by the user. That is an extraordinary hit rate.

The specific problems that `str` has helped us diagnose include

- *mutations* — the dimension of the object is wrong, suggesting that earlier code that was used to construct or manipulate the object is incorrect;

- *misclassification* — the object or its parts have unexpected classes, so the default treatment is unexpected;

- *misattributions* — the object lacks dimensions or other attributes that the code expects; and

- *synecdoche* — the user expects the object to return a part but asks for the whole — especially common with fitted models.

1.3.2.1 Debugging in Batch

Using `str` and its ilk is straightforward when one has access to the prompt, but all too often debugging happens in functions, and all too often the error messages are not very informative. Then we want to be able to stop processing, and examine the state. This is what the `browser` and `recover` functions allow. If we insert `browser()` in our function, then execution stops at that point, and we are able to visit the function's environment to examine the objects therein. If we type

```
> options(error = recover)
```

then when an error is detected, instead of stopping and dropping to the prompt, R will stop and offer a menu of frames for further examination. Further information can be found in a range of sources, including Jones et al. (2009), which provides simple worked examples of debugging.

1.3.3   Object-Oriented Programming

Object-oriented programming (OOP) is a type of programming that is built around several core values:

- encapsulation, which means that information can be hidden;

- inheritance, which means that code can be easily shared among objects; and

- polymorphism, which means that procedures can accept or return objects of more than one type.

Craig (2007) offers a readable introduction.

In order to introduce OOP in the context of R, we briefly turn to a programming conundrum: how to design a computer language that scales easily. Consider the operation of `print`ing an object that might be a number, or a model, or a matrix. Each of those three different classes of objects would require different code to print. Writing a single function that prints all possible classes would be clumsy and it would be challenging to extend in any unified way for new functionality. Therefore writing separate functions for each different class of object is necessary. However, what should all these different species of print functions be called? It would be tedious to have to recall a different function name for printing every different class of object. In base R alone, the number of different versions of `print` is

```
> length(apropos("^print."))
```

```
[1] 95
```

N.B.: Of course, that is just for the version of R installed on the second author's computer, which is

```
> sessionInfo()[[1]]$version.string
```

```
[1] "R version 2.15.2 (2012-10-26)"
```

So, we want many functions that can all be called using the same interface. OOP can solve this problem by polymorphism. However, even then we do not want to write endless links of `if` − `else` to handle the different scenarios. R provides different kinds of OOP facilities to solve this problem, both in the base language and via contributed packages. Here we cover the simplest implementation: S3 classes.

1.3.4   S3 Classes

Simple object-oriented programming is baked into R by means of the so-called S3 classes[1]. S3 classes provide a flexible and lightweight OO facility. They lack

---

[1] The name refers to the version of S in which the classes were introduced.

the protective infrastructure of more formal frameworks, such as S4 classes[2], but they are significantly easier to deploy. We consider Chambers (2008) to repay careful study on the topic, and note that the draft 'R Language Definition' is also useful reading. However, the underlying C code is the definitive resource.

Methods that are constructed for S3 classes rely on some base functions, called *generic* functions, which are used like building blocks. An example of a generic function is `print`.

```
> print
```

```
function (x, ...)
UseMethod("print")
<bytecode: 0x4319c70>
<environment: namespace:base>
```

When generic functions are called, they examine the class of the first argument. The function then calls a specific method that is identified as the function name and the class name, separated by a period. If the object class is, for example, *ml_g_fit*, then the print method will be `print.ml_g_fit`. If no method exists for the particular class, then the default function is called instead, specifically, `print.default`. Hence, S3 classes solve the function naming problem introduced above, by polymorphism. The programmer can write a method that is specific to the class, but all the user needs to do is use the generic function.

If an object has more than one class, as ours will, then the generic function uses the method for the first match among the listed classes. This is how S3 classes provide a kind of inheritance. Consider the following object.

```
> ordinal <- ordered(c(1,2,3))
> class(ordinal)
```

```
[1] "ordered" "factor"
```

When we try to print the ordinal object, the generic function will use the `print` method for the *ordered* class, if one exists (which it does not), or then that for the *factor* class if one exists, or the `default` method. So, the object class identifies the options in the preferred order, and here, the class of ordered objects kind-of inherits the methods of the factor class. We now print the ordinal object.

```
> ordinal
```

```
[1] 1 2 3
Levels: 1 < 2 < 3
```

---

[2]Ditto.

The reason that we are being a little coy about inheritance in S3 classes is that the inheritance is working at the object level, as opposed to the class level. It is quite possible to create mutant objects that do not follow the assumed phylogenetic structure, *viz.*

```
> class(ordinal) <- "ordered"
> ordinal

[1] 1 2 3
attr(,"levels")
[1] "1" "2" "3"
attr(,"class")
[1] "ordered"
```

Note that the object has lost the *factor* classification, and hence also its connection to `print.factor`. This can happen because the objects can be and mostly are created by means other than a formal constructor, and the object's class can be altered freely.

We can determine what methods have been written for a class using the `methods` function, as follows.

```
> methods(class = "ordered")

[1] Ops.ordered           Summary.ordered
[3] as.data.frame.ordered relevel.ordered*

   Non-visible functions are asterisked
```

This output shows us that there is no `print` method for the *ordered* class. We can also check to see what methods have been written for generic functions. Here, for example, we determine what classes have bespoke methods for the `nobs` generic function.

```
> methods(nobs)

[1] nobs.default* nobs.glm*     nobs.lm*      nobs.logLik*
[5] nobs.nls*

   Non-visible functions are asterisked
```

We mention in passing that all of these objects are flagged as being non-visible, meaning that they are not exported in the namespace. We can still study them, however, using `getAnywhere`.

```
> getAnywhere(nobs.lm)
```

```
A single object matching 'nobs.lm' was found
It was found in the following places
  registered S3 method for nobs from namespace stats
  namespace:stats
with value

function (object, ...)
if (!is.null(w <- object$weights)) sum(w != 0) else
NROW(object$residuals)
<bytecode: 0x4e097c0>
<environment: namespace:stats>
```

We close this section with an example of S3 programming. We invent a simple object, called `item`, and we suppose that `item` should have two classes, namely *thing1* and *thing2*, and that *thing2* should inherit some functionality from *thing1*.

```
> item <- 1:2
> class(item) <- c("thing2","thing1")
```

We write a simple `print` method for class *thing1*.

```
> print.thing1 <- function(x, ...) {
+    cat("inherits from Thing 1.\n")
+ }
```

Now we write a `print` method for class *thing2* that will also call the print method that we wrote for *thing1*. In this way we can share the same piece of code between the different classes of things. The `NextMethod` function, when called without arguments, will match the method called (here, `print`) with the next class in the object's class attribute (here, *thing1*). So it will be as though we called `print.thing1(x)`. `NextMethod` prevents endless recursion.

```
> print.thing2 <- function(x, ...) {
+    cat("Thing 2 ")
+    NextMethod()
+ }
```

We test our code by printing the item.

```
> item
```

```
Thing 2 inherits from Thing 1.
```

Now, the key to inheritance is that we want to be able to call this code for items of class *thing2*, but perhaps also *thing3*.

```
> print.thing3 <- function(x, ...) {
+    cat("Thing 3 also ")
+    NextMethod()
+ }
```

The following code shows the effect.

```
> another.item <- 1:3
> class(another.item) <- c("thing3","thing1")
> another.item
```

```
Thing 3 also inherits from Thing 1.
```

1.4   Making R Packages

We now cover what is arguably the key to R's great success as a statistical language: the R package system. This system is both a protocol and set of tools for sharing R functions and data. We do not intend to provide an exhaustive description of the system, but rather to provide enough information that readers will feel confident in constructing their own package, and about why it might be a good idea to do so.

A well-constructed R package is an unparalleled device for sharing code. Ideally it contains functions, documentation, datasets, and other information. Ideally the receiver will be able to install the R package and then run code examples that are provided in the help files that will execute on data that are installed from the package, and thus reproduce results that are either important in their own right, or assist the extension of the code to cover cases that are valuable to the user.

An R package is simply a compressed collection of files that conforms to a reasonably general directory structure and content. It is important to know that R packages conform to a wide range of specificity of detail, so that although it is possible to construct sophisticated combinations of R code and source code with various facilities, even a single function and appropriate documentation can comprise a package.

It is also important to distinguish between packages that are constructed for local use and packages that are intended for submission to a community such as the Comprehensive R Archive Network (CRAN). Packages intended for submission to CRAN are required to pass a rigorous battery of tests that should be replicated and checked on a local machine before submission. Packages for local use are subject only to local considerations; however, most of the provided tests for CRAN are useful regardless. This is not true of all the tests, for example, CRAN prefers that examples run quickly, which is less of an issue for local packages.

1.4.1 Building a Package

Building a package is a two-step process: first, creating a collection of directories and files according to a protocol, most often simply by using the `package.skeleton` function within R; and second, compiling the directories and files into a compressed binary archive using R tools in a command line interface (CLI).

A package starts out as a collection of directories and files contained within a single directory, which we will call the *package directory*. The most common structure has a `DESCRIPTION` file at the top level of the package directory, along with an `R` directory for the R functions, a `data` directory to hold any compressed R objects, and a `man` directory that holds the correctly formatted help files. That is all.

Other items, such as an `INDEX` file, are generated on the fly. A `NAMESPACE` file is also created automatically if one is not provided.

We do not have to create these directories and files ourselves. R can do that for us, using objects stored in memory. When we need to start making a package we invariably use the `package.skeleton` function, which constructs a basic and complete package framework using the objects in a nominated environment, for which the default is the global environment. The framework is complete in that all the needed files are present; however, the help files do require editing. That is, the easiest way to start building a package is to load the functions and the data into R, and then run the `package.skeleton` function. The latter will then construct the directories files that are needed, as described above. For example, the following call will create the infrastructure that is needed to build a package called *myPackage* that contains the objects `myFunction` and `myData`, and to save that infrastructure in the current working directory.

```
package.skeleton(name = "myPackage",
                 list(myFunction, myData),
                 path = ".")
```

Once the infrastructure is in place, building an R package requires a number of software tools. Covering all the options is beyond the scope of this book, but we will say that the needed tools are likely installed by default if your operating system is related to Unix, readily available as part of additional components of the Macintosh OS X, and available online as the executable archive `Rtools.exe` for the Windows family of operating systems. We will now assume that you have obtained and installed the needed software.

Whether we have constructed the package framework by hand, or using the `package.skeleton` function, we next create the package using the command-line interface (CLI) appropriate to the operating system. Readers who are unfamiliar with this step would benefit from background reading.

First, be sure that the files are up to date and reflect the desired contents. Pay particular attention to the help files and the `DESCRIPTION` file. Open the CLI and navigate to the directory that holds the package framework. We want the working directory to contain the package directory as identified above, say, *myPackage*. Then at the prompt for the CLI (here, $, which is the default prompt for the bash shell) type

```
$ R CMD build myPackage
```

R will construct the package. In order to see the options, type

```
$ R CMD build --help
```

If we want to include any other files with our package, then we add an `inst` directory, whose contents are included in package archive. This can be useful for including other scripts or raw datasets. We do this after running `package.skeleton`.

1.4.2 Testing

Before we distribute the package, or submit it to CRAN, we need to test it. The test will provide numerous checks of internal consistency and completeness. Imagine that the package that we have constructed is `myPackage_1.0.0.tar.gz` (note that the version number is determined by the contents of the `DESCRIPTION` file at the time of package construction). We run the tests by means of the following code.

```
$ R CMD check myPackage_1.0.0.tar.gz
```

We reiterate that these tests are not mandatory unless the package is destined to be submitted to CRAN, but they are a very useful device regardless. Very detailed output is provided on the test outcomes. We then iterate through the process of editing the package files to correct the errors, rebuilding the package, and applying the tests, until satisfied.

1.4.3 Installation

Finally, we can install the package into our R library using

```
$ R CMD install myPackage_1.0.0.tar.gz
```

The details of installation can be specific to your operating system, and particularly your permissions to write to various directories, so we omit further details.

1.5   Further Reading

There are many excellent books on programming. We mention particularly
Polya (1988), Michalewicz and Fogel (2010), Venables and Ripley (2000), Pace
(2012), Jones et al. (2009), and Chambers (2008).

1.6   Exercises

1. What is the relationship of a probability distribution function and
   a statistical model?

2. If matrix M is defined as

   ```
   > M <- matrix(c(3,4,6,8), nrow = 1)
   ```

   and matrix P is defined as

   ```
   > P <- matrix(c(3,4,6,8,4,8,4,7,2,2,5,4,4,7,5,2),
   +            ncol = 4)
   ```

   multiply M and P to have a 1×4 matrix, Q. What are the values
   in the vector Q?

3. Using R's PDF function, calculate the Poisson probabilities from 0
   through 10 given a mean value of 4.

4. Using R's pseudo-random number generator, generate 10 random
   Poisson variates given a Poisson mean value of 4.

5. (a) What is numerical optimization?
   (b) What is the distinction between a model and optimization al-
       gorithm?
   (c) Is there a universal optimization algorithm?
   (d) What are optimality conditions?
   (e) How does sensitivity analysis relate to numerical optimization?
   (f) What is the difference between discrete and continuous opti-
       mization?

6. Use `options(error = recover)` and `ls` to verify that `m` exists in
   the environment in which `variance.binomial` is being called in
   the example in Section 1.2.3.3, but not in the environment created
   within the `variance.binomial` execution.

2

Statistics and Likelihood-Based Estimation

2.1 Introduction

In the previous chapter we provided an overview of various R programming tools that will be needed when we start developing methods of model estimation. We also introduced the foremost probability and cumulative density functions that will be used in maximum likelihood estimation and simulation. Statistical modelling rests upon underlying probability functions, or mixtures of them, which are conceived to describe particular data situations. In this chapter we shall describe the relationship of data to probability and to likelihood, and show how these are in turn related to fitting and interpreting statistical models.

2.2 Statistical Models

Statistical models are used to describe a sample of data taken from a real or theoretical population. Statistical models can be described using one or more underlying probability distributions. The parameters of the distributions are estimated from the data, and may provide the basis for predicting additional data with the same distributional characteristics of the data being modeled. Models that can be defined in terms of a probability distribution having estimable parameters are called parametric models. We will focus our attention in this text on this type of model.

We note that non-parametric and semi- or partial-parametric models, as well as exact statistical models, have been constructed to deal with data that cannot be handled using normal parametric techniques. One set of non-parametric models is entirely, or in part, parameterized by smoothing splines. Statisticians generally turn to non-parametric model design when they have difficulty linearizing a continuous explanatory predictor so that it maintains an additive effect in an otherwise parametric model. Exact statistical models are used for dealing with sparse and unbalanced data. However, they demand huge amounts of computing power, and are thus limited in scope at this time.

Unless we specifically indicate otherwise, we henceforth are referring to parametric models when discussing statistical models in general.

Statistical models assume that the population elements of the variable of interest in the modelling task are randomly distributed according to a probability distribution, which is constrained by one or more parameters. It is natural to assume that each unit in the sample of data follows the same distribution as the population, so long as the sample is obtained by some unbiased method. Further, it is commonly assumed that the units of the sample are mutually independent — often with good reason — and that therefore the joint probability function of the sample can be obtained by multiplying the probability functions of each unit in the sample. Data that are clustered or correlated in some manner violate the distributional assumption of the probability function upon which the model is based. However, statisticians have derived a variety of adjustments so that unbiased parameter estimates may be calculated for correlated data, e.g., panel data.

Probability distributions may be either continuous or discrete, reflecting whether the variable of interest is continuous or discrete. When they are discrete, the probability distribution is commonly referred to as a probability mass function (PMF). When the variable being described in the model is continuous, we use the term probability density function, or PDF. In this text we will commonly refer to the *probability distribution function*, or *PDF*, by which we wish to refer to the PMF/PDF depending on the nature of the random variable.

Two additional defining characteristics of PMFs should also be mentioned. First, if we specify a random variable X as the event, place, thing, or person that is being described by a probability mass function, we may indicate the probability that X can take a specific value x as $f(x)$ or $f_{X=x}$. $f(x)$ is nonnegative for all real values of x.

Second, if X is described by a PMF, then the sum of all possible values of X is 1. That is, $\sum f(x_i) = 1$. When x is continuous, $\int f(x)dx = 1$. However, for continuous distributions, the probability of the random variable taking the value of any particular x is 0, unlike the probability for x in a discrete distribution.

Probability functions are often defined in terms of a location and one or more scale parameters. Many of the discrete distributions have only the location or mean parameter. Continuous distributions generally have one or two scale parameters in addition to the location parameter. Some discrete distributions, e.g., the negative binomial and beta-binomial distributions, have two parameters.

We mentioned earlier that statistical models are used to estimate the parameter(s) of the distribution to which the variable of interest belongs. This variable is typically referred to by statisticians as the *response variable* or *dependent variable*. If, for example, the response variable consists of independent count data, it is most likely assumed to be conditionally distributed as Poisson, having a specific mean value, λ. If the count data are correlated, then we might assume that the conditional distribution is negative binomial. The task of a Poisson statistical model, for example, is to estimate the mean parameter of Poisson distribution that characterizes the response term. Note that a mean or location parameter of Poisson or negative binomial models is represented by μ when being estimated as a generalized linear model (GLM). When not estimated as a GLM, their mean parameter is typically represented by λ. Exceptions exist in the literature, and which symbol is used in fact makes little difference as long as it is used consistently in the estimation process.

Above we say conditionally distributed because we mean that the data are distributed in this way given the parameters. That is, we want to allow for the possibility that the PDF parameters may be different for different observations in the data, and we construct a model to allow for it. For example, if we were to observe some data from the following distribution,

$$y_i \sim \mathcal{N}(\beta_0 + \beta_1 x_i, \sigma^2) \tag{2.1}$$

then we are not saying that the sample of y_i is normally distributed, but rather that it is normally distributed conditional on the x_i, β_0, β_1, and σ^2. Each unique observation of y has its own unique distribution, all of which are normal, and differ only in the parameters. This is a statistical model, and we can write this model more formally as follows:

$$f(Y_i = y_i | x_i, \beta_0, \beta_1, \sigma^2) = f_{\mathcal{N}}(y_i; \mu = \beta_0 + \beta_1 x_i, \sigma^2 = \sigma^2) \tag{2.2}$$

We are saying that the probability distribution of the random variable Y_i is normal, given that the other elements are known.

As we have seen, a key feature of a statistical model is that the distribution of the response variable, y, may be adjusted by one or more explanatory predictor variables, also called predictors, which we represent here by x. How the predictors relate to the response will become clear when we discuss the creation of synthetic models later in this book. The adjustment made by explanatory model predictors will affect the estimated mean parameter, and scale parameter(s) if appropriate. The predictors of two-parameter models may actually change the manner in which the adjusted response is distributed. This occurs more frequently with count data where predictors may result in the adjusted response being modeled as negative binomial rather than as Poisson. For single-parameter models the predictors do not affect which distribution is used for the model.

2.3 Maximum Likelihood Estimation

2.3.1 Process

Recall that the probabilities that are derived from a PDF are described by parameters. When we are modelling with data, we want to estimate the parameters of the model using the data. The parameters of a probability function are usually not directly estimated in statistical modelling. Instead, the conditioning of the PF is reversed. When the relationship of observations to parameters are reversed for a given probability function, statisticians refer to the function as a *likelihood function*. For a given probability distribution, we may write $f(y|\theta)$ where y represents the data and θ is the distribution parameter that produces y. Then the corresponding likelihood function is $L(\theta|y)$. The functional form is identical; all that changes is the conditioning. The probability refers to the probability of *data* conditional on *parameters*, whereas the likelihood refers to the likelihood of *parameters* conditional on *data*.

When models are estimated using maximum likelihood, the likelihood is transformed by the natural logarithm so that the contributions from each unit of the dataset are summed (under the assumption of conditional independence of the observations of the population), instead of being multiplied. This is because summing across values is numerically more stable than is multiplying across values. We will reserve $L(\theta|y)$ to refer to the log-likelihood of the parameters conditional on the data.

For an example we consider a Poisson model. The probability distribution for a single observation is

$$f_{Y=y}(y|\lambda) = \frac{\lambda^y e^{-\lambda}}{y!} \qquad (2.3)$$

where y is the response variable and λ is the mean or location parameter. The data are determined by the mean parameter via the PDF. A product sign would be placed in front of the probability function for an independent and identically distributed (iid) sample of observations.

The Poisson distribution belongs to a more general set of models termed the exponential family of distributions. We will provide more information about this family in Section 2.3.2.1. The probability functions of all member distributions can be recast in the form

$$f(\mathbf{y}|\theta, \phi) = \exp\left(\frac{y\theta - b(\theta)}{\alpha(\phi)} + c(y, \phi)\right) \qquad (2.4)$$

where θ is the link, $\alpha(\phi)$ is the scale, $b(\theta)$ is the cumulant, and $c(y, \phi)$ is the normalization function that guarantees that the PDF sums (or integrates) to 1. The derivative of the cumulant with respect to θ is the mean, and the second derivative is the variance of the distribution. This relationship is very useful in terms of modelling data.

The scale, $\alpha(\phi)$, is set at 1 for discrete probability distributions. Since the Poisson is a discrete distribution, we can drop the scale from the exponential equation. Parameterized so that the contribution of each observation to the overall PDF is given, we have

$$f(\mathbf{y}; \theta) = \prod_{i=1}^{n} \exp\left(y_i\theta_i - b(\theta_i) + c(y_i, \phi)\right) \qquad (2.5)$$

The Poisson distribution may be expressed in exponential family form as

$$f(\mathbf{y}; \theta) = \prod_{i=1}^{n} \exp\left(y_i \log(\mu_i) - \mu_i - \log(y_i!)\right) \tag{2.6}$$

The likelihood function is then the same:

$$L(\theta; \mathbf{y}) = \prod_{i=1}^{n} \exp\left(y_i \log(\mu_i) - \mu_i - \log(y_i!)\right) \tag{2.7}$$

As noted above, statisticians typically log both sides of the likelihood equation to obtain the log-likelihood.

$$\mathcal{L}(\theta; \mathbf{y}) = \sum_{i=1}^{n} y_i \log(\mu_i) - \mu_i - \log(y_i!) \tag{2.8}$$

Given the defining characteristics of the terms of the exponential family form, the Poisson link function is $\log(\mu)$ and the cumulant is μ. The inverse link function for members of the exponential family define the fitted value, symbolized as μ.

An alternative parameterization involves the linear predictor, $\mathbf{X}\beta$, which symbolizes the sum of the products of data and coefficients,

$$(\mathbf{X}\beta)_i = \beta_0 + \beta_1 x_{1i} + \beta_2 x_{2i} + \ldots + \beta_p x_{pi} \tag{2.9}$$

where $(\mathbf{X}\beta)_i$ is taken to mean the i-th row of the matrix $\mathbf{X}\beta$. In terms of the linear predictor, $\mu_i = (\mathbf{X}\beta)_i$. The Poisson log-likelihood, expressed in terms of the linear predictor, is then

$$\mathcal{L}(\theta; \mathbf{y}) = \sum_{i=1}^{n} y_i (\mathbf{X}\beta)_i - \exp(\mathbf{X}\beta)_i - \log(y_i!) \tag{2.10}$$

The quantity $\log(y_i!)$ may be computed efficiently in R using the `lfactorial` function

```
> y <- 5
> log(factorial(y))

[1] 4.787492

> lfactorial(y)

[1] 4.787492
```

A quick exploration shows that `lfactorial` is merely a wrapper for the `lgamma` function.

```
> lfactorial

function (x)
lgamma(x + 1)
<bytecode: 0xdd61ac>
<environment: namespace:base>
```

To obtain the coefficients or slopes for each of the predictors in the model, we take the first derivative of the log-likelihood function with respect to β, the parameter estimates, set it to 0, and solve. The function that is the derivative of the log-likelihood with regards to β is also called the gradient function or score function. For the Poisson distribution we have

$$\frac{\partial \mathcal{L}(\beta|y)}{\partial \beta} = \sum_{i=i}^{n}(y_i - \exp(\mathbf{X}\beta)_i)x_i \qquad (2.11)$$

Setting the above function to 0 yields what is referred to as the estimating equation for the Poisson model. The square matrix of second derivatives of the function with regards to each of the parameters is called the Hessian matrix.

$$\frac{\partial^2 \mathcal{L}(\beta|y)}{\partial \beta \partial \beta'} = -\sum_{i=i}^{n}(\exp(\mathbf{X}\beta)_i)x_i x_i' \qquad (2.12)$$

The negative of the inverse Hessian produces the estimated variance-covariance matrix of the parameter estimates of the model. Therefore the square root of the diagonal terms of the negative inverse Hessian matrix are large-sample estimates of the standard errors of the respective model coefficients.

The method of estimation described above is called *maximum likelihood estimation*. Other estimation methods exist, which we discuss in this volume. But the majority of statistics that we have displayed are relevant to other types of estimation. For example, estimation using Iteratively Re-weighted Least Squares (IRLS) is variety of maximum likelihood estimation. It is a shortcut made available due to the unique features of the exponential family, and which can be used to estimate parameters for models of that family. Since the Poisson model is a member of the exponential family, it may be estimated using IRLS techniques, which we describe in detail in Chapter 4. It may also be estimated using a full iterative Newton–Raphson type of algorithm, which can be expressed as the solution to

$$\beta_{j+1} = \beta_j - H^{-1}g \qquad (2.13)$$

where g is the gradient or first derivative of the log-likelihood function described above, H is the Hessian, and β_j are the coefficients.

Finally, it should be noted that the mean and variance of the Poisson distribution are then obtained by taking the first and second derivatives of the cumulant, respectively.

$$b'(\theta_i) = \frac{\partial b}{\partial \mu_i} \frac{\partial \mu_i}{\partial \theta_i} = 1 \times \mu_i = \mu_i \qquad (2.14)$$

and

$$b''(\theta_i) = \frac{\partial^2 b}{\partial \mu_i^2} \left(\frac{\partial \mu_i}{\partial \theta_i} \right)^2 + \frac{\partial b}{\partial \mu_i} \frac{\partial \mu_i^2}{\partial \theta_i^2} = 0 \times 1 + \mu_i \times 1 = \mu_i \qquad (2.15)$$

Finally, we mention that the method of estimation we have discussed thus far is based on a frequency interpretation of probability. The logic of this method is based on the notion that a statistical model represents a random sample of a greater population. The population may be factual or theoretical; e.g., all future instances or events. The method of estimation is conceptually based on the random sampling of the population an infinite number of times. We will discuss alternative viewpoints in Chapter 7.

2.3.2 Estimation

We now demonstrate maximum likelihood estimation of the single parameter of Watson's distribution, using R code. Recall from the previous chapter that the PDF is

$$f(x; \theta) = \frac{1 + \theta}{\theta \left(1 + \frac{x}{\theta}\right)^2} \qquad 0 < x \leq 1; \quad \theta > 0 \qquad (2.16)$$

This equation translates to the following log-likelihood.

$$\mathcal{L}(\theta; x) = \log(1 + \theta) - \log(\theta) - 2 \times \log\left(1 + \frac{x}{\theta}\right) \qquad 0 < x \leq 1; \quad \theta > 0 \quad (2.17)$$

In R, for a vector of data x, the function is as follows.

```
> jll.watson <- function(theta, x) {
+    sum(log(1 + theta) - log(theta) - 2*log(1 + x / theta))
+ }
```

We can maximize this function across θ a number of ways. We will use the optim function here, and we write a wrapper function for it to simplify our future usage. Our wrapper function is

```
> watson.fit <- function(x, ...) {
+    optim(0.5,
+          jll.watson,
+          x = x,
+          method = "Brent",
+          lower = 0, upper = 1,
+          control = list(fnscale= -1), ...)
+ }
```

A few points should be noted. First, we have only one parameter to estimate, so we select the option that calls the function `optimize`; this is `method = "Brent"`. This algorithm requires bounds to be provided for the parameter estimate. We call the wrapper as a function of some data as follows. The data are

```
> large.sample <- rep(1:10, 10)/20
```

and we fit the model using

```
> large.sample.fit <- watson.fit(large.sample)
```

   We can now examine the results of the model fit. The following quantities can be reported: `par` is the MLE for the parameter θ, `value` is the log-likelihood evaluated at the maximum, and the other three provide feedback on the execution. Both the `convergence` and `message` objects are of particular interest: a `convergence` of 0 reports that the algorithm has converged.

```
> large.sample.fit

$par
[1] 0.489014

$value
[1] 25.75886

$counts
function gradient
      NA       NA

$convergence
[1] 0

$message
NULL
```

We will cover how to obtain interval estimates in Section 2.4.

2.3.2.1 Exponential Family

The majority of parametric models currently employed for statistical modelling are based on the exponential family of distributions. These include such models as Gaussian, or normal, binomial, with links for logistic, probit, complementary log–log, and other related models, gamma, inverse Gaussian, Poisson, and negative binomial. Each of these models estimates the parameters of the underlying distribution based on the data at hand using a type of regression procedure or algorithm.

The basic one-parameter family of exponential models is referred to as *generalized linear models* (GLM), but two-parameter as well as three-parameter models exist as extensions of the continuous GLM distributions, as well as for binomial and negative binomial models. Mixtures of these models are also widely used, as well as truncated, censored, bivariate, and panel model varieties.

At the base of the exponential family is the generic exponential family probability function, commonly expressed as

$$f(y; \theta, \phi) = \exp \frac{y\theta - b(\theta)}{\alpha(\phi)} + c(y; \phi) \qquad (2.18)$$

where y is the response variable, θ is the canonical parameter or link function, $b(\theta)$ is called the cumulant, $\alpha(\phi)$ is called the scale parameter, which is set to 1 in discrete and count models, and $c(y, \phi)$ is a normalization term, used to guarantee that the probability function sums or integrates to unity. The negative binomial scale parameter has a different relationship with the other terms in the distribution compared to continuous distributions. We shall describe that difference later in the book.

The exponential family distribution is unique among distributions in that the first and second derivatives of the cumulant term, with respect to θ, yield the mean and variance functions, respectively.

It is important to note that models of this kind enjoy a host of fit and residual statistic capabilities, and are therefore favored by statisticians who are able to use them for their modelling projects. That is, it is reasonably straightforward to assess and compare the fits of GLMs to data; consequently, they are popular tools.

2.3.3 Properties

When certain regularity conditions are met, maximum likelihood estimates are characterized by having four properties, namely consistency, the asymptotic normality of parameter estimates, asymptotic efficiency, and invariance. Many times statisticians tend to ignore the regularity conditions upon which these properties are based; they are simply assumed to hold. However, ignoring the conditions can lead to biased and inefficient parameter estimates.

Regularity conditions include, but are not limited to

1. The elements or observations of the response variable, Y, are conditionally independent and identically distributed with a density of $f(y; \theta)$.
2. The likelihood function $L(\theta; y)$ is continuous over θ.
3. The first and second derivatives of the log-likelihood function are capable of being defined.
4. The Fisher information matrix is continuous as a function of the parameter. The Fisher information matrix cannot be equal to zero.

When these regularity conditions hold, then maximum likelihood estimators have the following properties.

Asymptotic consistency As the number of observations in a model goes to infinity, the ML estimator, $\hat{\theta}$, converges to its true value. That is, as $n \to \infty$, $|\theta - \hat{\theta}| \to 0$.

Asymptotic normality As the number of observations in a model goes to infinity, the ML estimator, $\hat{\theta}$, becomes normally distributed, and the covariance matrix becomes equal to the inverse Fisher or expected information. Even in models with relatively few observations, parameter estimates approach normality. For this reason it is possible to use the normal z-value to calculate confidence intervals.

Asymptotic efficiency As the number of observations in a model goes to infinity, the ML estimator, $\hat{\theta}$, has the smallest variance of any other estimator. An alternative way of expressing this is to state that an estimator is asymptotically efficient if it has the minimum asymptotic variance among competing asymptotically unbiased estimators. See Hilbe (2011).

Invariance The maximum likelihood estimator is invariant if it selects the parameter value providing the observed data with the largest possible likelihood of being the case. In addition, given that an estimator is in fact the MLE for a parameter, and that $g(\theta)$ is a monotonic transformation of θ, the MLE of $p = g(\theta)$ is $\hat{p} = g(\hat{\theta})$.

The first two properties are usually given by mathematical statisticians as the most important of the properties; the larger the number of observations, the more likely it is that the true parameter is being well approximated, and that the sampling distribution of this parameter estimate is normal.

Bias is the difference between the true and expected values of a parameter. An estimator is *unbiased* when the mean of its sampling distribution is equal to the true parameter, $E(\theta) = \theta$. There is no guarantee that any particular MLE will be unbiased, although some are, coincidentally.

The characteristics of inference from maximum likelihood estimation that we will use are based on large-sample asymptotics. When the model has only a relatively few observations, these characteristics or properties may break down, and adjustments need to be made for them. Of course, if we are able to obtain more data and it appears that the estimator is veering from the true parameter, or is becoming less normal, then there may be unaccounted-for bias in the data. In such a case, MLE may not be an appropriate method of estimation. Attempting to maximize a response term that is multi-modally distributed will also result in biased estimates, and in some cases in the inability of obtaining estimates at all.

Specification tests may be valuable for testing such possibilities, in particular Hausman specification tests evaluate whether an MLE estimator is efficient and consistent under the hypothesis that is being tested.

All in all though, maximum likelihood is a powerful estimation method, in particular for models with an exponential distribution base. It may still be considered as the standard method of estimation.

2.4 Interval Estimates

Here we cover two ways of producing large-sample interval estimates: Wald intervals and inversion of the likelihood ratio test.

2.4.1 Wald Intervals

Traditional (Wald-style) confidence intervals are calculated using the following formula

$$\hat{\beta}_j \pm z_{\alpha/2} \text{se}(\hat{\beta}_j) \tag{2.19}$$

where for $\alpha = 0.05$, corresponding to a 95% confidence interval, $z_{\alpha/2} = 1.96$. The assumption is that the sampling distribution of the coefficient estimate is normal. This assumption may be checked by graphical diagnostics.

An asymptotic estimate of the standard errors of the parameter estimates can be obtained from the call to **optim** by requesting retention of the Hessian matrix. To do this we simply need to add **hessian = TRUE** to the function call, as per the following example.

Here we create two samples, one large and one small, and fit the model to each. The standard error estimates provide feedback about the amount of data available in each sample. Note that these estimates are asymptotic, so they rely on the sample being large enough that the shape of the likelihood in the region of the optimum is appropriately quadratic.

```
> large.sample <- rep(1:10, 10)/20
> large.sample.fit <- watson.fit(large.sample, hessian = TRUE)
> large.sample.fit$par
```

```
[1] 0.489014
```

```
> (large.se <- sqrt(diag(solve(-large.sample.fit$hessian))))
```

```
[1] 0.1117381
```

The 95% Wald CI is then

```
> large.sample.fit$par + c(-1,1) * large.se
```

```
[1] 0.3772759 0.6007521
```

We now generate a much smaller sample, and fit the model to it, in order to observe the effect of sample size upon the certainty of the parameter estimates.

```
> small.sample <- rep(1:10, 1)/20
> small.sample.fit <- watson.fit(small.sample, hessian = TRUE)
> small.sample.fit$par
```

```
[1] 0.489014
```

```
> (small.se <- sqrt(diag(solve(-small.sample.fit$hessian))))
```

```
[1] 0.3533468
```

The 95% Wald CI is substantially wider, as follows.

```
> small.sample.fit$par + c(-1,1) * small.se
```

```
[1] 0.1356672 0.8423608
```

2.4.2   Inverting the LRT: Profile Likelihood

For models with few observations and models with unbalanced data, coefficients generally fail to approach normality in their distribution. In fact, most coefficient estimates are not distributed normally.

Due to the fact that the normality of coefficients cannot be assumed, many statisticians prefer to use the likelihood ratio test as a basis for assessing the statistical significance of predictors. The likelihood ratio test is defined as

$$-2 \left(\mathcal{L}_{\text{reduced}} - \mathcal{L}_{\text{full}} \right) \tag{2.20}$$

with $\mathcal{L}$ indicating the model log-likelihood, as before. The reduced model is the value of the log-likelihood with the predictor of interest dropped from the model or set to a value, usually 0 or 1, at which the contribution of the corresponding parameter estimate is negligible. Statistical significance is measured by the Chi2 distribution with 1 degree of freedom.

The LRT can be inverted to provide confidence intervals, which are in this case the range of values $\tilde{\beta}$ for which the null hypothesis that $\beta = \tilde{\beta}$ would *not* be rejected. The profiled confidence intervals do not rely on the assumption of normality of the parameter distribution; however, the use of the Chi2 distribution with 1 degree of freedom to provide the cutoff is still an asymptotic argument. The difference is that the Wald confidence intervals are based on a linear approximation to the true log-likelihood, whereas the LRT-based intervals use a quadratic approximation to the true log-likelihood (see e.g., Pawitan, 2001, §9.4).

We now demonstrate computing likelihood ratio intervals for estimates from Watson's distribution. First, we write a function to simplify calling the likelihood function for a range of values of θ.

```
> mll.watson <- function (x, data)
+    sapply(x, function(y) jll.watson(y, data))
```

Now we can evaluate the log-likelihood at the data and at this range of candidate parameter estimates.

```
> profiles <- data.frame(thetas = (1:1000)/1000)
> profiles$ll.large <- mll.watson(profiles$thetas, large.sample)
> profiles$ll.small <- mll.watson(profiles$thetas, small.sample)
```

Note that we don't need to refit the model at this point. We simply want to evaluate the log-likelihood at a large number of different potential parameter estimates.

In order to be able to superimpose summaries of the datasets, we subtract the maximum log-likelihood from each one. In doing so, we scale the log-likelihoods to each have a maximum of 0.

```
> for (i in 2:3)
+    profiles[,i] <- profiles[,i] - max(profiles[,i])
```

Finally, we can plot the two profiles on a single set of axes, and see what the interval estimates are, and also what effect the larger sample size has upon the symmetry and width of the intervals (Figure 2.1). We added a horizontal line at -1.92, which is the cutoff computed by halving the 0.95 quantile of the Chi2 distribution with 1 d.f.

```
> par(las = 1, mar = c(4,4,2,1))
> plot(ll.small ~ thetas, data = profiles,
+       type = "l", ylim = c(-4,0),
+       ylab = "Log-Likelihood (Scaled)",
+       xlab = expression(paste("Candidate Values of ", theta)))
> lines(ll.large ~ thetas, data = profiles, lty = 2)
> abline(h = -1.92)
> abline(v = large.sample.fit$par)
```

We see that the estimated 95% confidence interval for the small sample is approximately (0.16, 1.0) and for the large sample is about (0.33, 0.8). Clearly the increase in sample size has decreased the width considerably, and has made the interval more symmetric.

A two-parameter likelihood can be handled in the same way, producing profile contours instead of lines. A grid of candidate values for the two parameters is established, and the log-likelihood evaluated at each point on the grid. When there are more parameters than can be plotted, or one of the parameters is of primary import, then we need to find some way to ignore the other parameters.

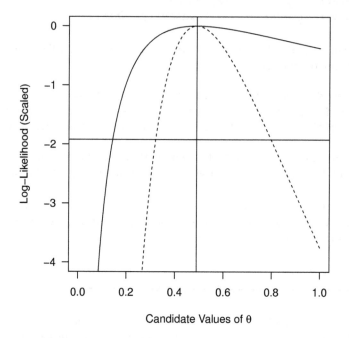

FIGURE 2.1
Profile log-likelihoods for a small sample (solid line) and a large sample
(dashed line) for θ, the parameter of Watson's distribution. A horizontal line
is added at -1.92 and a vertical line at the MLE.

2.4.3   Nuisance Parameters

The process of developing interval estimates for parameters in models that
have multiple parameters is more complicated. The reason for this complication is that the estimates of the parameters can and often do interact with one
another. Specifically, our best guess as to an interval estimate of one parameter can depend on the estimate of the other parameters. Further, sometimes
the other parameters are of interest to the analyst — and sometimes they are
not of interest — they are *nuisance parameters*.

Obtaining Wald interval estimates is the same as above: extraction of the
Hessian matrix and the subsequent computation. Analysts will often check the
correlation matrix of estimates to be sure that the estimates are reasonably
independent, and if they are satisfied as to this point, they will then calculate
and interpret the Wald interval estimates for each parameter of interest. We
will not consider Wald intervals further in this section.

LR-intervals are another matter. It is tempting to simply evaluate the log-
likelihood at a range of candidate values for the parameter or parameters of
interest, and replace the other parameters with their unconditional MLEs.

However, doing so ignores the potential relationships that the parameters could well have with one another. That is, we cannot ignore the potential effects that candidate values of the parameters of interest might have upon the estimates of the other parameters unless we are certain that the parameters are independent. Doing so will underestimate the size of the confidence regions and quite possibly produce misleading shapes of the confidence regions.

For example, consider the gamma distribution, with PDF as follows.

$$f(x) = \frac{1}{s^a \Gamma(a)} x^{a-1} e^{-x/s} \tag{2.21}$$

We might be interested in an estimate of parameter a (the shape parameter) but not s (the scale). We start with ML estimation as before,

```
> set.seed(1234)
> gamma.sample <- rgamma(1000, scale = 4, shape = 2)
> jll.gamma <- function(params, data) {
+    sum(dgamma(data,
+              scale = params[2],
+              shape = params[1],
+              log = TRUE))
+ }
> gamma.fit <- function(data, ...) {
+    optim(c(2,2),
+          jll.gamma,
+          data = data,
+          control = list(fnscale = -1), ...)
+ }
> test <- gamma.fit(gamma.sample, hessian = TRUE)
> test

$par
[1] 2.095975 3.869300

$value
[1] -2963.146

$counts
function gradient
      53       NA

$convergence
[1] 0

$message
NULL
```

```
$hessian
           [,1]       [,2]
[1,] -608.3007 -258.4447
[2,] -258.4447 -139.9754
```

Now we want to write a new version of the log-likelihood that allows us to specify the shape parameter. In order to construct a profile, we must be able to fix the parameter that we are interested in, and then maximize the log-likelihood across the rest of the parameters. We can think of this as being a conditional log-likelihood: the log-likelihood of s, the nuisance parameter, conditional on a given value of a, the parameter of interest, for which we want to compute the profile.

```
> jll.gamma.shape <- function(params, alpha, data) {
+    sum(dgamma(data, scale = params, shape = alpha, log = TRUE))
+ }
> gamma.fit.shape <- function(alpha, data, ...) {
+    optim(1,
+          jll.gamma.shape,
+          data = data,
+          alpha = alpha,
+          method = "Brent",
+          lower = 0, upper = 10,
+          control = list(fnscale = -1), ...)$value
+ }
```

Now we can fit the conditional log-likelihood for a range of pre-determined values of a and obtain the MLE. Here we use the convenient sapply command.

```
> gamma.profile <-
+    data.frame(candidates = seq(1.9, 2.3, length.out = 100))
> gamma.profile$profile.right <-
+    sapply(gamma.profile$candidates,
+           gamma.fit.shape,
+           data = gamma.sample)
```

Before we plot the result, we calculate the profile with the nuisance parameter fixed at its earlier MLE, and scale the two profiles to have maximum log-likelihood 0:

```
> gamma.fit.shape.wrong <- function(alpha, data, mle) {
+    sum(dgamma(data, scale = mle, shape = alpha, log = TRUE))
+ }
> gamma.profile$profile.wrong <-
+    sapply(gamma.profile$candidates,
+           gamma.fit.shape.wrong,
+           data = gamma.sample,
```

```
+                    mle = test$par[2])
> for (i in 2:3)
+    gamma.profile[,i] <- gamma.profile[,i] -
+        max(gamma.profile[,i])
```

Finally, Figure 2.2 is created with the following code.

```
> par(las = 1, mar = c(4,4,2,1))
> plot(profile.right ~ candidates, type="l",
+       data = gamma.profile, ylim = c(-2, 0),
+       ylab = "Log-Likelihood (Scaled)",
+       xlab = "Candidate Values of a")
> lines(profile.wrong ~ candidates, type="l",
+       data = gamma.profile,
+        lty = 2)
> abline(h = -1.92)
> abline(v = test$par[1])
```

We see that the effect of ignoring the nuisance parameter by setting it at its conditional MLE is considerable.

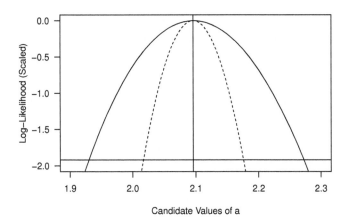

FIGURE 2.2
Profile log-likelihoods refitting the model (solid line) and not refitting (dashed line) for a, the shape parameter of the gamma distribution. A horizontal line is added at -1.92 and a vertical line at the MLE.

2.5   Simulation for Fun and Profit

We now introduce some tools that will be useful for exploring the properties of MLEs.

2.5.1   Pseudo-Random Number Generators

For models that are based on the exponential family of distributions, as well as on various other distributions, we may use the inverse transformation of a probability function to generate random variates belonging to that distribution. Such random numbers are created using the uniform distribution, which is `runif` in R. The `runif` function, which provides the probability distribution

$$f(x) = \frac{1}{b-a}; \quad a < x \le b \tag{2.22}$$

can be used to create random number generators from both discrete as well as continuous probability distributions, e.g., Poisson, binomial, and gamma.

It should be made clear that when discussing random numbers and random number generators, we nearly always are referring to pseudo-random numbers. Computers, unless given specialist software, do not calculate truly random numbers; rather, they compute thoroughly deterministic values based on the computer clock. However, they do appear to be random, and for practical purposes may be regarded as random.

The notion of a seed is central to understanding the pseudo-random values that are produced by the `runif` function. A seed is given to algorithms that use pseudo-random number generators so that identical results may be produced. This is sometimes useful in textbooks so that readers can obtain the same values as displayed in the text.

The seed is a series of numbers which are given to the `set.seed` function, which in turn commences generating numbers based on a pre-defined series of numbers. The value of the seed can be thought of as setting the place where numbers are deterministically read from a master list of numbers. When a seed value is not specifically given to `runif`, the algorithm sets the seed value based on the computer clock. In either case, the numbers are constrained to uniformly fit within the range 0 to 1.

We demonstrate the deterministic nature of calculating pseudo-random numbers using `runif` to generate two 2-number vectors of pseudo-random numbers, r1 and r2, with an initial seed number of 2468. We then use the same seed value to generate 4 pseudo-random numbers.

```
> testnum <- 2
> set.seed(2468)
> r1 <- runif(testnum)
```

```
> r2 <- runif(testnum)
> c(r1, r2)
```

[1] 0.4625722 0.5741233 0.3747903 0.9006981

```
> testnum <- 4
> set.seed(2468)
> (r12 <- runif(testnum))
```

[1] 0.4625722 0.5741233 0.3747903 0.9006981

Note that the concatenation of r1 and r2 is identical to r12. In fact, any time we employ that specific seed number, the values generated from using runif will be identical.

We shall drop the *pseudo-* prefix from random number generators for ease of reading. However, it is important to keep the actual operations in mind. Using the computer clock, whose values change each 1000-th or less of a second, we can produce what can be regarded as random numbers for nearly all practical applications.

Note that R already has plentiful operational random number generators for a range of probability distributions, and they are more stable and efficient than those that we will write here. See ?set.seed for generous documentation.

We shall first construct exponential and Chi2 random number generators based on runif to demonstrate the logic of developing random number generators. Thereafter we shall develop a Poisson random number generator. Together these should demonstrate how generators can be produced.

We develop an exponential RNG using the inverse transform method. Basically, the inverse transform method relies on the fact that for any probability function f, a random draw from its cumulative distribution F will always be uniformly distributed. Hence, drawing a uniformly distributed random number, say $q; q \in (0,1)$, and computing $x = F^{-1}(q)$ results in a random number x that is distributed as f.

```
> rndexp <- function(obs = 10000, shape = 3) {
+    xe <- -(shape)*log(runif(obs))
+    return(xe)
+ }
```

We use this function as follows.

```
> set.seed(1)
> rndexp(10, 3)
```

[1] 3.9783234 2.9655852 1.6713765 0.2888462 4.8031903 0.3214541
[7] 0.1707421 1.2429222 1.3903282 8.3522223

A Chi2 (with q degrees of freedom) random-number generator can be constructed using the normal (Gaussian) inverse CDF by summing the squares of q independent random normal variates. We can do this using a `for` loop, or an `apply` function, but it is faster to generate all the needed numbers in one call, convert them into an $n \times q$ matrix, and then use the efficient `rowSums` function to sum them.

```
> rndchi2 <- function(obs = 10000, dof = 3) {
+     z2 <- matrix(qnorm(runif(obs*dof))^2, nrow = obs)
+     return(rowSums(z2))
+ }
```

We use this function as follows.

```
> set.seed(1)
> rndchi2(10, 2)
```

```
[1] 1.0656127 0.9686650 0.2713120 1.8552262 1.2433986 1.6191096
[7] 2.8764449 5.9542855 0.2018375 2.9545210
```

Random generation of a Poisson random variable is a trickier proposition because the Poisson is discrete. Here we use an algorithm from Hilbe and Linde-Zwirble (1995).

```
> rndpoi  <- function(mu) {
+     g <- exp(-mu)
+     em <- -1
+     t <- 1
+     while(t > g) {
+         em <- em + 1
+         t <- t * runif(1)
+     }
+     return(floor(em + 0.5))
+ }
```

```
> set.seed(1)
> rndpoi(4)
```

```
[1] 4
```

Note that in order to create multiple random numbers, we might wrap this function in a `for` loop. However, more efficient approaches are possible. We now adapt the algorithm to use vectorized processing.

```
> rndpoi  <- function(obs = 50000, mu = 4) {
+     g <- exp(-mu)
+     em <- rep(-1, obs)
```

```
+    t <- rep(1, obs)
+    while(any(t > g)) {
+       em <- em + (t > g)
+       t[t > g] <- t[t > g] * runif(sum(t > g))
+    }
+    return(floor(em + 0.5))
+ }
```

An example of using the above code is

```
> set.seed(1)
> xp <- rndpoi(10000, 4)
> str(xp)

 num [1:10000] 1 6 4 5 1 3 4 6 11 1 ...

> mean(xp)

[1] 4.0028

> var(xp)

[1] 4.066999
```

2.6   Exercises

1. What is the difference between a probability model and a statistical model?

2. Construct a function that will generate pseudo-random binomial numbers in a similar manner as Poisson function discussed in the chapter.

3. What is the difference between the observed and expected information matrix? Which is preferred for calculating the standard errors for non-canonical GLM models? Why?

4. Maximize the following equation with respect to x: $2x^2 - 4x + 1$.

5. (Challenging): Create a generic function to calculate profile likelihood confidence intervals following (a) a glm logistic model, (b) a glm Poisson model, (c) a glm negative binomial model, (d) any glm model.

3

Ordinary Regression

3.1 Introduction

Linear regression and its generalization, the linear model, are in very common use in statistics. For example, Jennrich (1984) wrote, "I have long been a proponent of the following unified field theory for statistics: *Almost all of statistics is linear regression, and most of what is left over is non-linear regression.*" This is hardly surprising when we consider that linear regression focuses on estimating the first derivative of relationships between variables, that is, rates of change. The most common uses to which the linear regression model is put are

1. to enable prediction of a random variable at specific combinations of other variables;

2. to estimate the effect of one or more variables upon a random variable; and

3. to nominate a subset of variables that is most influential upon a random variable.

Linear regression provides a statistical answer to the question of how a target variable (usually called the *response* or *dependent* variable) is related to one or more other variables (usually called the *predictor* or independent variables). Linear regression both estimates and assesses the strength of the statistical patterns of covariation. However, it makes no comment on the causal strength of any pattern that it identifies.

The algebraic expression of the linear regression model for one predictor variable and one response variable is

$$y_i = \beta_0 + \beta_1 \times x_i + \epsilon_i \tag{3.1}$$

where y_i is the value of the response variable for the i-th observation, x_i and ϵ_i are similarly the predictor variable and the error respectively, and β_0 and β_1 are the unknown intercept and slope of the relationship between the random variables x and y.

In order to deploy the model to satisfy any of these uses noted above, we need estimates of the unknown parameters β_0 and β_1, and for some of the uses we also need estimates of other quantities. Furthermore, different assumptions must be made about the model and the data for these different uses; we detail these assumptions below.

3.2   Least-Squares Regression

The challenge of determining estimates for the parameters, conditional on data, can be framed as an optimization problem. For least-squares regression, we are interested in finding the values of the parameters that minimize the sum of the squared residuals, where the residuals are defined as the differences between the observed values of y and the predicted values of y, called $\hat{y}$.

Exact solutions are available for least-squares linear regression, but our ultimate goal is to develop models for which no exact solutions exist. Therefore we treat least-squares linear regression in this manner as an introduction.

$$\min_{\beta_0, \beta_1} \sum_{i=1}^{n} (y_i - (\beta_0 + \beta_1 x_i))^2 \tag{3.2}$$

For example, consider the following observations, for which least-squares optimization is decidedly unnecessary.

```
> y <- c(3, 5, 7)
> x <- c(1, 2, 3)
```

We can write the objective function as a function in R, and use the powerful `optim` function to minimize the objective function across its first argument, which may be of any length. So, the least-squares objective function for obtaining estimates of β_0 and β_1 can be written in R as

```
> least.squares <- function(p, x, y) {
+    sum((y - (p[1] + p[2] * x))^2)
+ }
```

where x is the predictor variable, y is the response variable, and p is the vector of parameters.

We need to choose a starting point for the optimization routine, and we would like to be sure that the function can be evaluated at the starting point that we choose. So we choose some plausible values (here, $\beta_0 = 0$ and $\beta_1 = 0$) and run a brief test, as follows.

```
> start.searching.here <- c(intercept = 0, slope = 0)
> least.squares(start.searching.here, x, y)
```

```
[1] 83
```

We see that the function can be evaluated at that particular combination of parameter estimates and the response and predictor variables. This test is important as it provides direct feedback as to whether or not optim is likely to succeed. The least-squares estimates can then be obtained by the following call.

```
> optim(par = start.searching.here,
+        fn = least.squares,
+        x = x, y = y)$par

intercept      slope
0.9991065 2.0003141
```

We see that the estimate of the intercept is close to 1 and the estimate of the slope is close to 2. We now briefly describe the arguments that we have used for our call to optim.

- par is a vector that presents the starting point for the search. The dimension of the space to be searched is equal to the length of the vector. If the values are labeled, as here, then the labels are passed through optim to the output.

- fn is the function to be minimized. The function will be minimized across its (possibly multidimensional) first argument.

- x and y are the other arguments that we need to resolve the value of fn. Note that these arguments were declared when the least.squares function was created. This is a useful example of how the ... argument is used: optim accepts arguments that do not match its formal arguments and passes them to fn.

As is well known, the values that minimize the objective function (3.2) also have a closed-form expression; $\hat{\beta}_1 = \frac{SS_{xy}}{SS_{xx}}$ and $\hat{\beta}_0 = \bar{y} - \hat{\beta}_1\bar{x}$, where SS_{ab} refers to the sum of squares of variables a and b: $SS_{ab} = \sum(a_i - \bar{a})(b_i - \bar{b})$. Equivalently, in R, we can use

```
> mean(y) - cov(x,y) / var(x) * mean(x)    # Beta 0

[1] 1

> cov(x,y) / var(x)                        # Beta 1

[1] 2
```

Minimizing the sums of squares, also called the L2 norm, is a popular approach to obtaining estimates for the unknown parameters. However, other objective functions are also used, for example, minimizing the sum of the absolute values of the residuals (called the L1 norm), or minimizing the maximum of the absolute values of the residuals (called the L∞ norm). These alternative functions will often lead to different estimates, and the estimates will have different properties. For example, parameters that are estimated by minimizing the L1 norm are less affected by remote observations, or outliers. However, we do not expect that the estimates from the different objective functions will be particularly different for these data.

In R, minimizing the sum of absolute values of the residuals, which corresponds to minimizing the L1 norm, can be done as follows,

```
> L1.obj <- function(p, x, y) {
+    sum(abs(y - (p[1] + p[2] * x)))
+ }
> optim(c(0,0), L1.obj, x=x, y=y)$par

[1] 0.9999998 2.0000000
```

and minimizing the maximum absolute value of the residuals, which corresponds to minimizing the L∞ norm, is

```
> Linf.obj <- function(p, x, y) {
+    max(abs(y - (p[1] + p[2] * x)))
+ }
> optim(start.searching.here, Linf.obj, x=x, y=y)$par

intercept      slope
0.9999999 2.0000000
```

and as we suspected, for these data, the differences are negligible. The `optim` function is detailed earlier in this section.

3.2.1   Properties

The estimates that minimize the objective function (3.2) are, by definition, least-squares estimates, regardless of any other assumptions, the origins of the data, the validity of the model, and so on. These estimates can be used to solve the first challenge above, which is to predict values of y conditional on x. No further assumptions are required.

However, unless assumptions are made, the estimates lack statistical content. This point is sufficiently important to bear restating: in order for parameter estimates to have statistical content, certain specific assumptions must be made, and the assumptions must be checked. We describe the assumptions and the relevant diagnostics in this section, and provide an example of their use and checking in Section 3.2.4. Note that we present the assumptions in a specific sequence. The statistical properties of the estimates grow as we add more assumptions.

First assumption

If we assume that the x values are fixed and known, and that the functional form of the model (3.1) is correct, then the least-squares estimates are unbiased (see, e.g., Casella and Berger, 1990). We can check whether this assumption is reasonable by examining a scatterplot of the estimated residuals against the fitted values, perhaps augmented with a smooth line. If there is no substantial pattern to the average of the estimated residuals, then we have some justification for the assumption.

Second assumption

We further assume that the y observations have constant but unknown conditional variance. Note that if the conditional variance is not constant, then the quality of information that we have about the parameters varies depending on the values of the predictors, and this dependence is not captured by the model. The chance of being misled by a simple statistic is high. The assumption of constant conditional variance can be assessed by examining a scatterplot of the square root of the absolute value of the *standardized studentized* residuals against the fitted values, ideally augmented with a smooth line. If there is no trend to the smooth line, then it is reasonable to assume constant variance for the residuals.

Following Hardin and Hilbe (2007) we define *standardized* residuals as those that have had their variance standardized to take into account the correlation between y and $\hat{y}$; specifically, the residual is multiplied by $(1-h_i)^{-1/2}$ where h_i is the i-th diagonal element of the hat matrix (see Section 3.2.2). Also, we define *studentized* residuals as those that have been divided by an estimate of the unknown scale factor, here estimated by $\hat{\sigma}$. We will provide more variations on residuals in Chapter 4.

Third assumption

If we also assume that the observations are conditionally independent, then the least-squares estimates have the smallest variance among all unbiased linear estimates. Since the estimates were computed by minimizing the residual variation, this outcome should not be particularly surprising. The assumption of conditional independence is harder to check definitively. Checking this assumption appropriately will usually require the use of some knowledge about the design of the data collection. Generally the analyst will use information about the design to guide the choice of the types of dependence to check. For example, if groups of the observations have similar origin, clustering, then it may be worth checking for intra-group correlation, and if the data have a time stamp then checking for autocorrelation is an important consideration.

Fourth condition

This point is mentioned because it is relevant here, although it resides more naturally with Section 3.3. It is not really an assumption about the conditional distribution of the response variable, as such. Least-squares estimates can be expressed as sums of conditionally independent random variables, so the estimates are subject to the Central Limit Theorem. The interested reader can learn more from Huber (1981, Theorem 2.3, Chapter 7), Demidenko (2004, §13.1.1), and DasGupta (2008, Theorem 5.3 and Example 5.1). Consequently, asymptotically, the estimates are normally distributed. This observation can be used to justify an assumption of normality for the parameter estimates, which can in turn be used to construct interval estimates and hypothesis tests. However, the assumption of conditionally normal errors, as in Section 3.3, is more commonly used. We discuss this point further in Section 3.4.8.4.

Fifth assumption

Finally, we assume that the sample is representative of the population for which we wish to make inference. This assumption is often unstated, although it is usually checked, even if just at an intuitive level. This is the assumption that leads us to explore summary statistics of the sample, to assess outliers, and to focus attention on the possible effects of the sample design upon the outcome. If the sampling process does not permit the collection of a sample that represents the population, then inference will fail. We may interpret outliers or unusual patterns in the data in this light, and update our model, or we may conclude that the observations are erroneous.

3.2.2   Matrix Representation

Representing the linear model as we have done above (3.1) is straightforward when only a small number of variables is involved, and we are only interested in obtaining parameter estimates. However, that representation gets messier when we want more information from the model, or the model gets larger. Then, carefully selected matrices provide a compact and convenient way of representing the model and some important results about it.

Following the usual approach, we represent the response variable as an $n \times 1$ vector $\mathbf{Y}$, the predictor variables (including the intercept) as an $n \times p$ matrix $\mathbf{X}$, the parameters as a $p \times 1$ vector $\boldsymbol{\beta}$, and the residuals as an $n \times 1$ vector $\boldsymbol{\epsilon}$. Then the linear model is

$$\mathbf{Y} = \mathbf{X}\boldsymbol{\beta} + \boldsymbol{\epsilon} \tag{3.3}$$

Given this model formulation, it turns out that the least-squares estimates $\hat{\boldsymbol{\beta}}$ can be obtained in closed form using matrix manipulation as follows (see, among others, Weisberg, 2005).

$$\hat{\boldsymbol{\beta}} = (\mathbf{X}'\mathbf{X})^{-1}\mathbf{X}'\mathbf{Y} \tag{3.4}$$

We can demonstrate this solution with our toy example:

```
> X <- as.matrix(cbind(1, x))
> (beta.hat <- solve(t(X) %*% X) %*% t(X) %*% y)

    [,1]
      1
x     2
```

The parentheses surrounding the second statement are shorthand for "evaluate this expression and print the returned object." The estimate of the intercept is 1, and of the slope is 2. We created the *model matrix* (also called the *design matrix*) by binding a column of 1's with the predictor variable x. Note that R automatically repeated the value 1 as many times as was needed to match the length of x. We then used four matrix-specific functions;

1. %*%, which performs matrix multiplication,

2. t, which returns the transpose of a matrix, and

3. solve, which we used here to provide the inverse of $\mathbf{X'X}$.

There are other applications for solve; the reader should see the help file to learn more about them. More general presentations of these solutions do not require $\mathbf{X}$ to be of full rank, because $\hat{\beta}$ can be computed without explicitly inverting $\mathbf{X'X}$. See Section 3.2.3.

Other than a compact representation of the model and the least-squares solutions, the adoption of matrices here provides convenient representations of other quantities from the model that may be useful. For example, the variance of $\hat{\beta}$ can be written

$$\text{Var}\left(\hat{\beta}\right) = \sigma^2 \left(\mathbf{X'X}\right)^{-1} \tag{3.5}$$

This representation is a convenient and useful expression. The estimates can be computed in R as follows. We have to compute the variance of the residuals σ^2, which requires calculation of the fitted values, here denoted y.hat, by applying Equation (3.3).

```
> (y.hat <- X %*% beta.hat)

      [,1]
[1,]    3
[2,]    5
[3,]    7

> (sigma.2 <- as.numeric(var(y - y.hat)))

[1] 0

> (vcov.beta.hat <- sigma.2 * solve(crossprod(X)))
```

```
       X
     0 0
   x 0 0
```

Here we have replaced the `t(X) %*% X` by `crossprod(X)`, which efficiently computes $\mathbf{X'X}$ when given $\mathbf{X}$ as its only argument.

These results are as expected — the residual variance is negligible; our model was a perfect fit to the data. This rare opportunity should be enjoyed. Ordinarily, such a contingency will more likely signal an error of logic or modelling.

We can also write the variance of the predicted values and the residuals using the convenient matrix representation. We note that, from (3.3) and (3.4),

$$\hat{\mathbf{Y}} = \mathbf{X}(\mathbf{X'X})^{-1}\mathbf{X'Y} \tag{3.6}$$

then we take (as is traditionally done) $\mathbf{H} = \mathbf{X}(\mathbf{X'X})^{-1}\mathbf{X'}$, so $\hat{\mathbf{Y}} = \mathbf{HY}$ and $\hat{\epsilon} = (\mathbf{1} - \mathbf{H})\mathbf{Y}$. It can then be easily shown that, under the model,

$$\text{Var}(\hat{\mathbf{Y}}) = \sigma^2\mathbf{H} \tag{3.7}$$

and

$$\text{Var}(\hat{\epsilon}) = \sigma^2(1 - \mathbf{H}) \tag{3.8}$$

$\mathbf{H}$ is called the Hat matrix, and we can obtain it inefficiently (see later in this section for the preferred approach) as follows

```
> X.hat <- X %*% solve(t(X) %*% X) %*% t(X)
```

and then use it to compute the variances of the fitted values and the residuals,

```
> (var.y.hat <- sigma.2 * diag(X.hat))
```

```
[1] 0 0 0
```

```
> (var.e.hat <- sigma.2 * (1 - diag(X.hat)))
```

```
[1] 0 0 0
```

which are negligible for this model and these data, as expected.

Finally, we note that the variance of the residuals is of course only an estimate of the variance of the errors, and is therefore subject to the same kinds of uncertainty as are our parameter estimates. Analysts will usually use the t-distribution as a template for describing the uncertainty of parameter estimates to accommodate this additional uncertainty.

3.2.3 QR Decomposition

The approach to determining the parameter estimates using the matrix representation presented in Section 3.2.2 is never used in professionally developed software. This is because the solution relies on inversion of a matrix $(\mathbf{X}'\mathbf{X})$ that may be large and ill-conditioned, which means that the parameter estimates may be difficult to obtain accurately. Direct inversion leads to slow and possible inaccurate estimation.

At the time of writing, R uses QR decomposition to obtain representations of the model matrix that are easier to work with and numerically more stable. S uses QR decomposition by default, although Choleski decomposition and singular-value decomposition are also available (Chambers, 1992b)

QR decomposition relies on the decomposition of the model matrix $\mathbf{X}$ into two components, labeled $\mathbf{Q}$ and $\mathbf{R}$. The operations that are necessary to obtain the parameter estimates from $\mathbf{Q}$ and $\mathbf{R}$ enjoy greater numerical stability than those required to resolve (3.4). The difference between the naive and the decomposed approach is analogous to the difference between two different ways of computing the sums of squares, namely that

$$SS_{xx} = \sum x_i^2 - \frac{\left(\sum x_i\right)^2}{n} \tag{3.9}$$

is efficient because it requires only one loop across the data, but

$$SS_{xx} = \sum (x_i - \bar{x})^2 \tag{3.10}$$

is numerically more stable because it sums the squares of differences, rather than computing the difference between two (possibly large) quantities. (See Chan et al. (1983) for a useful discussion.)

QR decomposition works as follows. When $\mathbf{X}$ is of full rank and size $n \times p$, we wish to find $\mathbf{Q}$ ($n \times p$) and $\mathbf{R}$ ($p \times p$) such that

1. $\mathbf{X} = \mathbf{Q}\mathbf{R}$

2. $\mathbf{Q}'\mathbf{Q} = \mathbf{I}$

3. $\mathbf{R}$ is upper triangular, i.e., all the entries below the diagonal of $\mathbf{R}$ are 0.

One way to find such a decomposition of $\mathbf{X}$ is via the Householder transformation (Algorithm 5.2.1, Golub and Van Loan, 1996). Briefly, the Householder transformation involves a sequence of p matrix pre-multiplications upon $\mathbf{X}$, each of which targets a specific column of $\mathbf{X}$. The effect of each pre-multiplication is to make the below-diagonal elements of the corresponding column of $\mathbf{X}$ into 0.

R provides a `qr` function that uses the DQRDC code from LINPACK (Dongarra et al., 1979), and functions `qr.R` and `qr.Q` to extract the $\mathbf{R}$ and $\mathbf{Q}$ matrices, respectively. We show the function and verify its operation for our example design matrix below.

```
> (xR <- qr.R(qr(X)))

                    x
[1,] -1.732051 -3.464102
[2,]  0.000000 -1.414214

> (xQ <- qr.Q(qr(X)))

            [,1]           [,2]
[1,] -0.5773503  7.071068e-01
[2,] -0.5773503  2.775558e-16
[3,] -0.5773503 -7.071068e-01

> xQ %*% xR

     x
[1,] 1 1
[2,] 1 2
[3,] 1 3
```

The advantage of working with a QR decomposition of $\mathbf{X}$ is that the estimates $\hat{\boldsymbol{\beta}}$ may now be computed by solving

$$\mathbf{R}\hat{\boldsymbol{\beta}} = \mathbf{Q}'\mathbf{Y} \tag{3.11}$$

This equation can easily be computed by *backsolving*, because R has upper triangular structure. Backsolving is an algorithm that involves finding the unknown values for $\hat{\boldsymbol{\beta}}$ one element at a time, exploiting the upper-triangular nature of R. Backsolving is more stable and more efficient than generic matrix inversion. We now demonstrate backsolving using our example dataset and model.

```
> Y <- matrix(c(3,5,7), nrow=3)
> xR

                    x
[1,] -1.732051 -3.464102
[2,]  0.000000 -1.414214
```

Note the structural zero in the lower left corner.

```
> t(xQ) %*% Y

            [,1]
[1,] -8.660254
[2,] -2.828427
```

We see that we can directly calculate the estimate of β_1 as $-2.82/-1.41 = 2$, and conditional on that estimate we can compute the estimate of β_0 as 1, without a potentially messy matrix inversion. Note that R also provides a `backsolve` function, viz.:

```
> backsolve(xR, t(xQ) %*% Y)
```

```
        [,1]
[1,]      1
[2,]      2
```

These results agree exactly with our formulation above.

We will also need to obtain the hat matrix in order to standardize the residuals. The hat matrix is $\mathbf{X}(\mathbf{X}'\mathbf{X})^{-1}\mathbf{X}'$, but equivalently can be expressed as $\mathbf{Q}\mathbf{Q}'$, which is faster to compute and does not require inversion of $\mathbf{X}'\mathbf{X}$. Also, for standardizing the residuals we only need the diagonal elements of the hat matrix, so computing and storing the whole thing is unnecessary, and may be cumbersome for large datasets. In R,

```
> (hat.values <- diag(crossprod(t(xQ))))
```

```
[1] 0.8333333 0.3333333 0.8333333
```

The algorithm used when $\mathbf{X}$ is not of full rank is summarized by Chambers (1992b). This scenario is relatively common, for example, when the predictor variables include one or more factors.

We conclude this section by noting that QR decomposition also provides a more stable way of computing the estimated covariance matrix for the parameter estimates. Recall that from Equation (3.5), we would compute the covariance as $\sigma^2(\mathbf{X}'\mathbf{X})^{-1}$. However, we know that $\mathbf{X} = \mathbf{Q}\mathbf{R}$ and that $\mathbf{Q}'\mathbf{Q} = \mathbf{I}$, so the covariance can also be written as

$$\text{Var}\left(\hat{\boldsymbol{\beta}}\right) = \sigma^2(\mathbf{R}'\mathbf{R})^{-1} \tag{3.12}$$

which is more efficient to calculate than (3.5) because $\mathbf{R}$ is upper-triangular in structure, and smaller than $\mathbf{X}$.

3.2.4 Example

We now use some tree measurement data to demonstrate the least-squares parameter estimation approach. These data are taken from a forest inventory of the University of Idaho Experimental Forest, in the Upper Flat Creek stand. The data are measures of tree species and the tree diameter at 1.37 m from the ground for all sampled trees, and also tree heights in a purposively selected subsample. The sample design was a systematic grid of variable-radius plots, but this is not relevant to our example, so we will not use the information.

The measurement data are provided as an object named ufc in the *msme* package that accompanies this book. We will construct a model that we can use to predict the height of a tree as a function of its diameter. We will ignore measurement error in the diameters, although it is known to exist. That is, we will treat the diameters as being known and fixed, which is standard in the discipline.

```
> library(msme)
> data(ufc)
```

We will also sweep out all the missing values using na.omit, with the caveats presented in Section 3.4.2.

```
> ufc <- na.omit(ufc)
```

We will use the matrix representation to obtain our least-squares estimates. Our response variable is ufc$height.m, and our predictor variable is ufc$dbh.cm. First, we form the response vector and the model matrix.

```
> X <- cbind(1, ufc$dbh.cm)
> Y <- ufc$height.m
```

We now obtain the QR decomposition of **X**.

```
> xR <- qr.R(qr(X))
> xQ <- qr.Q(qr(X))
```

Finally, our least-squares parameter estimates are obtained using backsolve, and the estimated covariance matrix is computed as per Equation (3.12).

```
> (beta.hat <- backsolve(xR, t(xQ) %*% Y))

           [,1]
[1,] 12.6757004
[2,]  0.3125935

> y.hat <- X %*% beta.hat
> (sigma.2 <- as.numeric(var(Y - y.hat)))

[1] 24.35307

> (vcov.beta.hat <- sigma.2 * solve(crossprod(xR)))

            [,1]           [,2]
[1,]  0.317353252 -0.0069997721
[2,] -0.006999772  0.0001920922
```

The large-sample estimates of the standard errors are then

```
> sqrt(diag(vcov.beta.hat))
```

[1] 0.56334115 0.01385973

At this point we have the estimators that minimize the residual variance. However, we do not know whether the estimators are unbiased, or even whether bias means anything in this context, and we will not know whether the covariance estimate of the parameters is meaningful, until we examine the relevant diagnostics. For these, we need the raw residuals and the standardized studentized residuals, which are the raw residuals divided by their estimated standard errors. See Equation (3.8) for the estimated variance of the residuals.

```
> e.hat <- Y - y.hat
> e.hat.ss <- e.hat /
+    sqrt(sigma.2 * (1 - diag(crossprod(t(xQ)))))
```

It cannot hurt to check the result of standardization.

```
> var(e.hat)
```

```
         [,1]
[1,] 24.35307
```

```
> var(e.hat.ss)
```

```
         [,1]
[1,] 1.006163
```

We are now in a position to produce the two graphical diagnostics that inform us about the first two assumptions (Figure 3.1). We adopt Hadley Wickham's *ggplot2* package for constructing our graphical diagnostics (Wickham, 2009). Here we deliberately omit plotting the default standard error regions, as they cannot be interpreted in the intuitively appealing way as limits for the underlying curve.

```
> qplot(y.hat, e.hat,
+        ylab = "Residuals", xlab = "Fitted Values") +
+    geom_abline(intercept = 0, slope = 0) +
+    geom_smooth(aes(x = y.hat, y = e.hat), se = FALSE)

> qplot(y.hat, abs(sqrt(e.hat.ss)),
+        ylab = "Standardized Studentized Residuals",
+        xlab = "Fitted Values") +
+    geom_smooth(aes(x = y.hat, y = abs(sqrt(e.hat.ss))),
+        se = FALSE)
```

The left panel of the figure suggests that the choice of a straight line as the model of the relationship between height and diameter might be mistaken. There seems to be substantial curvature, which would also have a realistic biological interpretation — physics would suggest that tree shapes must be

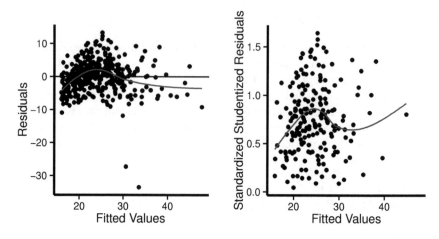

FIGURE 3.1
Diagnostic graphs for the least-squares fit of height against diameter using the tree measurement data from `ufc`.

such that height and diameter cannot be linearly related. This figure also shows the practical import of the assumptions trying to determine whether the slope and intercept are unbiased makes little sense when the model fails to capture important features of the relationship.

A similar conclusion may be drawn from the right panel of Figure 3.1: the conditional variance does not seem to be constant, although it is less worrisome in this instance as the deviation from the desired pattern is less.

Overall we would conclude that we need a more sophisticated model. Construction of such a model is deferred until a later section.

3.3   Maximum-Likelihood Regression

If we are willing to make a more stringent assumption about the relationship between the data and the model then we will realize further benefits for our estimates. Previously, we assumed that the observations are conditionally independent with identical variance and that the model form is correct. We may also assume that the residuals are normally distributed. If we do, then it is natural to think of maximum likelihood in this context, as we have finally brought enough assumptions to bear, and as described in Section 2.3.3, maximum likelihood estimates have desirable statistical properties.

The model for our data is now

$$y_i \stackrel{d}{=} \mathcal{N}(\beta_0 + \beta_1 x_i, \sigma^2) \tag{3.13}$$

Note that there are some important changes compared with the previous model.

First, we have written a completely specified joint PDF for the data. In order to fit this model we will, in theory, need to obtain parameter estimates for all the unknown parameters. Previously, in least-squares regression, we did not need to estimate σ^2 unless we were interested in obtaining estimates of the standard errors of our parameter estimates. Here, σ^2 is an integral part of the model. It turns out in this case that point estimates of the other parameters can be obtained without estimating σ^2, because of the structure of the log likelihood, but that is a consequence of the special nature of this particular model. Often we will be required to find estimates for all the parameters in the model, even the ones that we are not interested in interpreting. The latter will be called *ancillary parameters*. They are sometimes referred to as nuisance parameters, but we prefer to avoid the value judgment.

Second, we have specified the conditional distribution of the response variable. The model specifies that the response variable will be conditionally normally distributed. This is a useful assumption that, if demonstrated to be reasonable, will bring more powerful properties to our estimates. We will need to check this additional assumption as part of the fitting process. Note that if we fail to check such assumptions, then any claims that we make about the model that are derived from the calculated log-likelihood are unsupported!

Recall that maximum likelihood estimation is another optimization problem, like least-squares estimation, just with a different objective function. We now write a function that will evaluate the log-likelihood as a function of the data and the parameters.

```
> jll_normal <- function(p, x, y) {
+   sum(dnorm(y, p[1] + p[2] * x, p[3], log = TRUE))
+ }
```

The joint log-likelihood is a function of three parameters and two random variables. The three parameters (passed as argument p) are the intercept β_0, the slope β_1, and the scale parameter σ. The two random variables are the predictor variable x and the response variable y, respectively.

As before, we use optim to find the values of the parameters that maximize this function, conditional on the data.

```
> optim(par = c(intercept = 0, slope = 0, sigma = 1),
+       fn = jll_normal,
+       control = list(fnscale = -1),
+       x = x, y = y)$par

   intercept        slope        sigma
1.000000e+00 2.000000e+00 3.705307e-12
```

We find the same values for the intercept and slope of x as we did earlier for the OLS method of estimation. However, an estimate of the scale parameter is also displayed, given as 3.705e-12, which here is the machine equivalent of 0.

Our use of `optim` has introduced a new argument that we should explain. The `control` argument allows us to pass a `list` of arguments that can be used to tune the optimization. The alteration that we make here is to multiply the output of the function by −1, thus converting the minimization problem into a maximization problem. See Section 3.2 for an explanation of the other arguments.

3.4   Infrastructure

We now wrap the MLE engine in a function to simplify its deployment, and we will add some fitting infrastructure.

3.4.1   Easing Model Specification

R provides a formula class that enables the straightforward communication of certain kinds of models. Objects of this class are particularly useful for communicating linear predictors. We can use the `formula` object, in conjunction with a dataframe that contains the fitting data, to communicate the response variable and the linear predictor in R as follows.

```
ml_g <- function(formula, data) {
  mf <- model.frame(formula, data)
  y <- model.response(mf, "numeric")
  X <- model.matrix(formula, data = data)
  ...
}
```

This small section of code provides us with powerful model-handling capabilities. The `model.frame` function uses the `formula` and `data` objects to create a dataframe that contains and organizes the necessary pieces for fitting the model. We extract the response variable from `data` by using the `model.response` function.

We create the model matrix from the formula and data objects by using the `model.matrix` function. This approach will handle a wide range of different model specifications, for example, models that include interactions between predictor variables or transformations of predictor variables can all be communicated by the formula object and will be represented in the model matrix. Furthermore, `model.matrix` automatically creates the necessary dummy variables to represent any factors under one of a range of different types of contrasts.

Having touched upon factors, we now need to mention how they are handled. Inappropriate handling of factors in regression models leads to problems with parameter estimation. S handles factors in linear models by using devices called *contrasts*. The choice of contrast to use is guided by the intended application of the model. The default contrasts in R are

```
> options()$contrasts
```

```
        unordered             ordered
  "contr.treatment"        "contr.poly"
```

Treatment contrasts, which are used for categorical factors, are formed by setting the estimate of the first level of the factor to zero. Polynomial contrasts, used by R for ordinal variables, are formed so that the individual coefficients can be interpreted as values of orthogonal polynomials assuming that the levels of the factor are equally spaced numeric values (Chambers and Hastie, 1992).

By the end, we have the response variable y and the model matrix X ready for deployment in the optimizer. We mention in passing that *formula* class objects also handle specification of non-linear models. We will tackle this usage later. See Chambers and Hastie (1992) for further reading.

The big advantage of providing a model matrix and response variable to the optimizer is that a single piece of code can then be used to fit a very wide array of different and useful models. The joint log-likelihood can now be declared as follows. We will assume that the model parameters will be the p regression parameters followed by the standard deviation of the residuals.

```
> jll_normal <- function(params, X, y) {
+    p <- length(params)
+    beta <- params[-p]
+    sigma <- params[p]
+    linpred <- X %*% beta
+    sum(dnorm(y, mean = linpred, sd = sigma, log = TRUE))
+ }
```

This objective function can now be used for any linear predictor that we care to declare, including factors, interactions, splines, and so on — any model that can be decomposed into a linear model can now be fit, at least in theory, by maximizing this objective function.

3.4.2 Missing Data

As we noted in Chapter 1, one of our programming values is to keep our code as modular as is reasonable. There are many different reasons that observations may have missing variables, and those different reasons each motivate different responses for the analyst. Trying to distinguish between these contingencies, and write suitable code for them, creates an unreasonable amount of complexity for our function. We prefer to make the analyst directly responsible for handling missing data, so our function will test the data that will be used for the model and stop if missingness is detected. However, we will only apply our test to the data that are being used for the model; hence, we only test the response variable and the model matrix. We will include the following line.

```
if (any(is.na(cbind(y, X)))) stop("Some data are missing.")
```

3.4.3   Link Function

The objective function that we introduced in Section 3.4.1 provides functionality for a wide range of different linear models. However, it also has a weakness: evaluation in part of the parameter space will result in nonsensical results. The PDF for the normal distribution is undefined if the scale is negative. The performance of our optimizer will improve if we can avoid this kind of behavior. One way to constrain the parameter estimate is by introducing box constraints upon it, and `optim` permits these constraints if we use the appropriate argument. However, it is more instructive to introduce a transformation of the relevant parameter: a link function. We will replace the scale σ with a new scale $\exp(\sigma)$, which is always greater than zero. All we need to do is be sure that we interpret the parameter estimate correctly, which is *as the natural logarithm of the quantity of interest*. This development results in the following objective function.

```
> jll_normal <- function(params, X, y) {
+     p <- length(params)
+     beta <- params[-p]
+     sigma <- exp(params[p])
+     linpred <- X %*% beta
+     sum(dnorm(y, mean = linpred, sd = sigma, log = TRUE))
+ }
```

In addition to protecting part of the parameter space, link functions can provide parameter estimates with different and sometimes useful interpretations, and models that are naturally constrained to make predictions within certain ranges. We will be using the link function principle extensively for the linear predictors as well as the scale in the next chapter.

3.4.4   Initializing the Search

A good starting point can make the difference between convergence at the optimum and failure to converge, or worse, convergence at a local optimum that is far from the global optimum. The objective function for maximum likelihood with normal errors is quadratic in shape, but nonetheless it is possible for an optimizer to get lost. Here we will take advantage of the least-squares formulation of the problem. This approach will be more useful still in subsequent chapters. Rather than use the purpose-written code that we wrote earlier, we will use R's own fitting functions. We will provide our response variable and model matrix, and let R know that the model matrix already has an intercept, by including `-1` in the formula. Note that this code cannot be run as-is; we report it here for discussion purposes and will include it in our function, to follow.

```
ls.reg <- lm(y ~ X - 1)
beta.hat.ls <- coef(ls.reg)
sigma.hat.ls <- sd(residuals(ls.reg))
start <- c(beta.hat.ls, sigma.hat.ls)
```

We acknowledge that using least-squares estimates as starting points for maximum-likelihood regression with normal errors seems absurd from one point of view. However, it enables us to focus on the problem that motivates the text (MLE) instead of the more quotidian issue of the selection of start points for individual cases.

3.4.5 Making Failure Informative

Now that we are taking responsibility for the maximization of the likelihood, it is necessary for us to interpret the output of the optimizer carefully. If the optimizer is not confident that it has reached the optimum, then interpreting the parameter estimates is a dangerous step. We will check the convergence condition that is reported by `optim` and stop all processing if the condition is not 0.

```
if (fit$convergence > 0) {
  print(fit)
  stop("optim failed to converge!")
}
```

3.4.6 Reporting Asymptotic SE and CI

The use of maximum-likelihood estimation provides estimators that have useful and desirable properties, such as asymptotic normality. Hence if we have obtained such estimates, it makes sense to report summary values such as the standard errors and confidence intervals of the estimates.

As noted in Section 2.4.1, the asymptotic estimate of the covariance matrix of the maximum-likelihood parameter estimates can be obtained as a function of the inverse of the Hessian matrix, evaluated at the maximum-likelihood estimate. The Hessian matrix can be returned directly by `optim`, which makes calculation of the covariance matrix straightforward. The following code chunk demonstrates how to ask `optim` to return the Hessian.

```
fit <- optim(start,
             ...,
             hessian = TRUE)
```

We next extract the parameter estimates from the object returned by `optim`.

```
beta.hat <- fit$par
```

The covariance matrix is the inverse of the negative hessian, and the standard errors are the square roots of the diagonal members of the covariance matrix.

```
se.beta.hat <- sqrt(diag(solve(-fit$hessian)))
```

Finally we gather the various statistics into a neat coefficient table, which we plan to use as a default reporting device. In keeping with statistical tradition, we will include the Z ratio, which is the parameter estimate divided by the estimate of the standard error.

```
zTable <- data.frame(Estimate = beta.hat,
                     SE = se.beta.hat,
                     Z = beta.hat / se.beta.hat,
                     LCL = beta.hat - 1.96 * se.beta.hat,
                     UCL = beta.hat + 1.96 * se.beta.hat)
rownames(zTable) <- c(colnames(X), "Log Sigma")
```

We now have all the pieces that we need to create the function that will fit a linear regression using maximum likelihood, assuming normal errors.

3.4.7   The Regression Function

Our development will proceed as follows: we will write a function that performs maximum-likelihood regression to fit a linear regression model assuming normal errors. This function will accept as arguments, among other things, a model specification (using the *formula* class) and a dataset (using the *dataframe* class), and will return an object that reports various elements of the fitting procedure, including the outcome. Some of the functions used herein will be explained in greater detail in the following text.

```
> ml_g <- function(formula, data) {
+
+ ### Prepare the data, relying on the formula class for
+ ### handling model specification
+   mf <- model.frame(formula, data)
+   y <- model.response(mf, "numeric")
+   X <- model.matrix(formula, data = data)
+
+ ### Check for missing data.  Stop if any.
```

```
+    if (any(is.na(cbind(y, X)))) stop("Some data are missing.")
+
+ ### Declare the joint log likelihood function
+    jll_normal <- function(params, X, y) {
+        p <- length(params)
+        beta <- params[-p]
+        sigma <- exp(params[p])
+        linpred <- X %*% beta
+        sum(dnorm(y, mean = linpred, sd = sigma, log = TRUE))
+    }
+
+ ### Initialize the search
+    ls.reg <- lm(y ~ X - 1)
+    beta.hat.ls <- coef(ls.reg)
+    sigma.hat.ls <- sd(residuals(ls.reg))
+    start <- c(beta.hat.ls, sigma.hat.ls)
+
+ ### Maximize the joint log likelihood
+    fit <- optim(start,
+                 jll_normal,
+                 X = X,
+                 y = y,
+                 control = list(
+                     fnscale = -1,
+                     maxit = 10000),
+                 hessian = TRUE
+                 )
+
+ ### Check for optim failure and report and stop
+    if (fit$convergence > 0) {
+        print(fit)
+        stop("optim failed to converge!")
+    }
+
+ ### Post-processing
+    beta.hat <- fit$par
+    se.beta.hat <- sqrt(diag(solve(-fit$hessian)))
+
+ ### Reporting
+    results <- list(fit = fit,
+                    X = X,
+                    y = y,
+                    call = match.call(),
+                    beta.hat = beta.hat,
+                    se.beta.hat = se.beta.hat,
```

```
+                     sigma.hat = exp(beta.hat[length(beta.hat)]))
+
+ ### Prepare for S3 deployment (see next Section!)
+    class(results) <- c("ml_g_fit","lm")
+    return(results)
+ }
```

This function is then used as follows.

```
> ufc.g.reg <- ml_g(height.m ~ dbh.cm, data = ufc)
```

Note the lack of output from this command. The outcome is an R object that
is called `ufc.g.reg`. We now write functions to examine and use the object.

3.4.8   S3 Classes

The final innovation that we wish to introduce is a protocol for conveniently
handling the checking of assumptions and the reporting of our model. Now
we will write a small collection of helper functions, called *methods*. In order
to simplify the use of these functions, we will create them according to a
specific object-oriented template. R provides more than two implementations
of object-oriented programming; here we use the straightforward S3 classes.

Our ultimate goal is to produce a compact summary of the model and
appropriate regression diagnostics. We will write a number of small helper
methods that build up to this result. First, notice that the `results` object
has been assigned two classes: *ml_g_fit* and *lm*. This allows us to use methods
that have been written for *lm*, and replace them selectively with our own.

3.4.8.1   Print

The first requirement is to print the object. We notice that the generic `print`
function has a method for objects of class *lm*.

```
> print.lm
```

```
function (x, digits = max(3, getOption("digits") - 3), ...)
{
    cat("\nCall:\n", deparse(x$call), "\n\n", sep = "")
    if (length(coef(x))) {
        cat("Coefficients:\n")
        print.default(format(coef(x), digits = digits),
            print.gap = 2, quote = FALSE)
    }
    else cat("No coefficients\n")
    cat("\n")
    invisible(x)
}
<environment: namespace:stats>
```

Our objects inherit class *lm*, so we can use this `print` function if we can provide the infrastructure that it requires. In order to deploy this function for objects of our class, we need those objects to inherit class *lm*, and according to the structure of the `print.lm` method, the object should contain a `call` object (note the `x$call`), and the class should have a `coef` method (note the `coef(x)`). Given these three conditions, we can simply use the `print` method written for the *lm* class. The first two conditions are fulfilled in the function that is developed in Section 3.4.7. We will therefore start by writing a method that returns the coefficients of the regression line. Recall that these are all but the last of the estimates.

```
> coef.ml_g_fit <- function(object, ...) {
+    object$beta.hat[-length(object$beta.hat)]
+ }
```

Now when we call `coef` on our object, this function is used automatically.

```
> coef(ufc.g.reg)
```

```
X(Intercept)       Xdbh.cm
  12.6770795     0.3125628
```

However, that's not all. Now, when we call the generic `print` function with our fitted model as the first argument, or equivalently simply call the object itself, we obtain the same summary of the model that is used for `lm`.

```
> ufc.g.reg
```

```
Call:
ml_g(formula = height.m ~ dbh.cm, data = ufc)

Coefficients:
X(Intercept)       Xdbh.cm
     12.6771        0.3126
```

3.4.8.2   Fitted Values

We will now develop diagnostic tools for assessing our assumptions. Assessing assumptions should precede examining the model. Our second method uses the first to return the fitted values for the model from the fitted model object, calculated by the matrix multiplication of the parameter estimates $\hat{\beta}$ and the model matrix $\mathbf{X}$.

```
> fitted.ml_g_fit <- function(object, ...) {
+    as.numeric(object$X %*% coef(object))
+ }
```

We can quickly assess the effect of this method using `str` as follows. We are looking at both the structure and content of the object; it should be a vector of about 400 observations that look like tree heights measured in meters! Note that this step may seem unnecessarily painstaking; however, it is our experience that a large proportion of errors in R programming can be detected and solved by simply scrutinizing the returned objects.

```
> str(fitted(ufc.g.reg))

 num [1:391] 24.9 27.7 28.9 23.9 24.6 ...
```

This object seems to accord with our expectations. Also, note that in our call to the function we only needed to call `fitted`, rather than `fitted.ml_g_fit`. This is a consequence of method dispatch.

3.4.8.3 Residuals

In this model, the raw residuals are simply the difference between the predicted values and the observations. In order to standardize the residuals, we will need to compute the diagonal values of the hat matrix. R provides a generic function called `hatvalues` that we can use as a basis for our own method. We will use the efficient QR decomposition to obtain the hat matrix (see Section 3.2.3).

```
> hatvalues.ml_g_fit <- function(model) {
+   tcrossprod(qr.Q(qr(model$X)))
+ }
```

Again, we check that the output accords with our expectations in terms of structure and content. Now we are looking for an $n \times n$ matrix of positive values all less than 1.

```
> str(hatvalues(ufc.g.reg))

 num [1:391, 1:391] 0.00261 0.00279 0.00287 0.00255 0.00259 ...
```

We can now report the residuals of the fitted model. Here we introduce some small complications. There are presently two kinds of residuals that we are interested in: the raw residuals, and the standardized studentized residuals. We will ask the user to select between these two types using an argument. Further, we will make sure that the user has selected one of these two types, by using the `match.arg` function on the chosen argument. Finally, we will assume that users are interested in raw residuals if they decline to make a choice. The method proceeds as follows.

```
> residuals.ml_g_fit <-
+     function(object, type = c("raw","ss"), ...) {
+     type <- match.arg(type)
+     e.hat <- object$y - fitted(object)
```

```
+     if (type == "ss") {
+         e.hat <- e.hat /
+             (object$sigma.hat * sqrt(1 - diag(hatvalues(object))))
+     }
+     return(e.hat)
+ }
```

Note that the method makes use of the two methods that we have already written: `fitted.ml_g_fit` and `hatvalues.ml_g_fit`. However, because of the S3 class dispatch mechanism, again we can omit the class name in the calls to the functions. As before, we check the structure and content of the returned object. We hope that the following call will produce a vector of about 400 numbers that look like residuals that have unit variance.

```
> ufc.g.res.s <- residuals(ufc.g.reg, type = "ss")
> str(ufc.g.res.s)

Named num [1:391] -0.887 1.081 0.217 -0.656 -0.417 ...
 - attr(*, "names")= chr [1:391] "2" "3" "5" "8" ...

> var(ufc.g.res.s)

[1] 1.008413
```

The output provides us with no concerns about the operation of the method.

3.4.8.4 Diagnostics

As already noted in Section 2.3.3, maximum-likelihood estimates have useful statistical properties. In order to justify the invocation of these properties for our estimators, we must check that the assumptions that we need to make are satisfied.

The assumptions that are required for maximum-likelihood estimates are more onerous than those for least-squares regression. As with least-squares regression, maximum likelihood regression with normal errors requires that the functional form be correct, and that the errors be independent and have constant variance. However, we also need to assume that the errors are normally distributed.

In order to check whether the assumption of normality is reasonable, we can examine a quantile–quantile plot of the standardized studentized residuals against the normal distribution. We mention in passing that in generalized linear models, which we will cover in the next two chapters, numerous different types of residuals are used for different purposes (see e.g., Hardin and Hilbe, 2007). In the normally distributed case, covered here, the raw residuals $y - \hat{y}$ are the same as the Pearson and the deviance residuals, which we cover in Chapter 4.

Even if the match between the distribution of the residuals and the normal

distribution is not particularly good, we can take some comfort from the effect
of the Central Limit Theorem, because the estimates are least-squares estimates. Note that we need to standardize the residuals before comparing them
with the normal quantiles because if the residuals are not standardized then
they do not have constant variance (see Equation 3.8), and as a consequence
should not be expected to collectively match a single normal distribution.
Studentizing is not required for this comparison.

As a general rule, the effects of the failures of assumptions to hold can be
explored using *simulation*. We provide some pointers and examples in Chapter 7. The parametric and non-parametric bootstrap are also very useful tools;
see e.g., Davison and Hinkley (1997).

We now have sufficient infrastructure to develop a graphical diagnostic tool
that we can use to assess the quality of the fit of the model to the data. We will
now construct a function that provides useful feedback on the assumptions
that we rely upon for our statistical inference. This method will produce a
collection of four scatterplots:

1. observations (y-axis) against fitted values (x-axis) to assess the overall utility of the model,

2. raw residuals (y-axis) against fitted values (x-axis) to assess the lack
   of fit of the mean function,

3. square roots of the absolute values of the standardized studentized
   residuals (y-axis) against fitted values (x-axis) to assess the con-
   stancy of the variance, and

4. quantiles of the standard normal distribution (y-axis) against the
   sorted standardized studentized residuals (x-axis) to assess the assumption of normally distributed errors.

As before, we are using Wickham's *ggplot2* package to provide the graphics, and *gridExtra* to provide the layout. This choice adds somewhat to the
necessary infrastructure of the plot command; however, the payoff in ease of
control is considerable.

```
> plot.ml_g_fit <- function(x, ...) {
+    require(ggplot2)
+    require(gridExtra)
+    e.hat <- residuals(x)
+    e.hat.ss <- residuals(x, type="ss")
+    y.hat <- fitted(x)
+    n <- nrow(x$X)
+    pp1 <- qplot(y.hat, x$y,                        ### Plot 1
+             ylab = "Observations", xlab = "Fitted Values") +
+          geom_abline(intercept = 0, slope = 1) +
+          geom_smooth(aes(x = y.hat, y = x$y), se = FALSE)
+    pp2 <- qplot(y.hat, e.hat,                      ### Plot 2
```

```
+                    ylab = "Residuals", xlab = "Fitted Values") +
+           geom_abline(intercept = 0, slope = 0) +
+           geom_smooth(aes(x = y.hat, y = e.hat), se = FALSE)
+     pp3 <- qplot(y.hat, abs(sqrt(e.hat.ss)),        ### Plot 3
+                    ylab = "Sqrt (Abs( Stand. Res.))",
+                    xlab = "Fitted Values") +
+           geom_smooth(aes(x = y.hat, y = abs(sqrt(e.hat.ss))),
+                    se = FALSE)
+     pp4 <- qplot(sort(e.hat.ss), qnorm((1:n)/(n+1)), ### Plot 4
+                    xlab = "Stand. Stud. Residuals",
+                    ylab = "Normal Quantiles")
+     grid.arrange(pp1, pp2, pp3, pp4, ncol=2)
+ }
```

This function is then used as follows, to create Figure 3.2.

```
> plot(ufc.g.reg)
```

We already know that we think that this model is a poor choice; we learned that from the diagnostics of the least-squares regression in Figure 3.1. We will try to improve upon the model shortly.

Other diagnostics may also be useful depending on the context of the problem, for example, the analyst may wish to plot leverage against studentized residuals, Cook's distances, etc. See ?influence for more information. Furthermore, the user may wish to check for suspected autocorrelation, if the data have a time component or a hierarchical structure. For these cases, generally we would advocate also fitting a model that accommodates the correlation (e.g., a panel model, see Chapter 6) and comparing the two either using a statistical test or, better, a summary statistic that captures the purpose of model fitting.

3.4.8.5 Metrics of Fit

We now consider functions that we can use to report aspects of the fitted model. For example, we may be interested in reporting the log likelihood of the fitted model, evaluated at the maximum likelihood parameter estimates. There is a generic function to do so: logLik. The value that we wish to report is the value of the optimized function, which is the **value** object of the **fit** object of the ml_g_fit object(!). The structure of the following method borrows heavily from the logLik.lm method, which we examined in the process of writing the book, using the getAnywhere function.

```
> logLik.ml_g_fit <- function(object, ...) {
+     val <- object$fit$value
+     attr(val, "nall") <- nrow(object$X)
+     attr(val, "nobs") <- nrow(object$X)
+     attr(val, "df") <- length(object$fit$par)
```

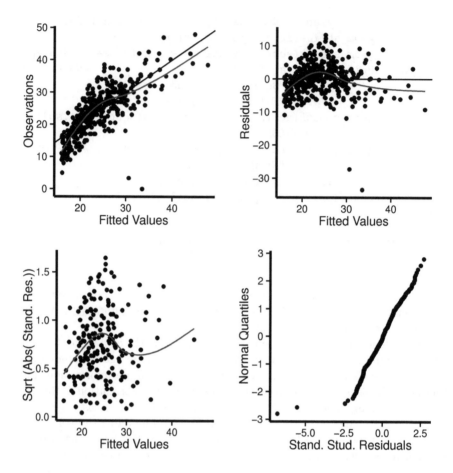

FIGURE 3.2
Four diagnostic graphs for the maximum-likelihood normal linear regression
of height against diameter using the tree measurement data from `ufc`.

```
+    class(val) <- "logLik"
+    val
+ }
```

The (maximized) log-likelihood for our model and data is then:

```
> logLik(ufc.g.reg)
```

```
'log Lik.' -1178.469 (df=3)
```

Notice that by mimicking the structure of the `logLik` methods, specifically
by adding relevant attributes and the appropriate class to the returned object,
we were able to take advantage of the existing `print.logLik` method.

We can now take advantage of the S3 class infrastructure. We may be interested in computing Akaike's Information Criterion (Akaike, 1973) for our model. We could certainly write our own function for this objective, but having carefully planned the structure of `logLik.ml_g_fit`, we can also take advantage of the existing `AIC` generic function, which simply relies on being able to call `logLik` upon the objects that are passed to it.

```
> AIC(ufc.g.reg)
```

```
[1] 2362.938
```

This rather trivial example demonstrates the utility of careful planning when programming in an object-oriented environment.

3.4.8.6 Presenting a Summary

Our final method will provide something similar to the printed output that we created in Section 3.4.6, and an object that returns some summary statistics for the model fit. Again, we can take example of an existing function by providing the infrastructure that is required for `print.summary.lm`, which we examined during the writing of this book using the very useful `getAnywhere` function.

```
> summary.ml_g_fit <- function(object, dig = 3, ...) {
+    zTable <- with(object,
+               data.frame(Estimate = beta.hat,
+                          SE = se.beta.hat,
+                          Z = beta.hat / se.beta.hat,
+                          LCL = beta.hat - 1.96 * se.beta.hat,
+                          UCL = beta.hat + 1.96 * se.beta.hat))
+    rownames(zTable) <- c(colnames(object$X), "Log Sigma")
+    p <- length(object$fit$par)
+    n <- nrow(object$X)
+    df <- c(p, n, p)
+    summ <- list(call = object$call,
+                 coefficients = zTable,
+                 df = df,
+                 residuals = residuals(object),
+                 aliased = rep(FALSE, p),
+                 sigma = object$sigma.hat)
+    class(summ) <- c("summary.ml_g_fit", "summary.lm")
+    return(summ)
+ }
```

Now our fitted model can be summarized in the following useful way.

```
> summary(ufc.g.reg)

Call:
ml_g(formula = height.m ~ dbh.cm, data = ufc)

Residuals:
    Min      1Q  Median      3Q     Max
-33.525  -2.863   0.132   2.851  13.320

Coefficients:
             Estimate       SE        Z      LCL     UCL
(Intercept) 12.67708  0.56271 22.52853 11.57416 13.780
dbh.cm       0.31256  0.01384 22.57707  0.28543  0.340
Log Sigma    1.59521  0.03577 44.60167  1.52511  1.665

Residual standard error: 4.929 on 391 degrees of freedom
```

We can check the collection of functions that we have written for the *ml_g_fit* class using the **methods** function, as follows.

```
> methods(class = "ml_g_fit")

[1] coef.ml_g_fit      fitted.ml_g_fit      hatvalues.ml_g_fit
[4] logLik.ml_g_fit    plot.ml_g_fit        residuals.ml_g_fit
[7] summary.ml_g_fit
```

Note firstly that we have not made any central register of the methods. R is simply picking them up based on the structure of the function names. Also note that the **print** function does not appear in this list. This is because it is inherited from the *lm* class.

The preceding development does not show the structural simplicity of the collection of functions that we have created. Briefly, **plot** requires **fitted** and **residuals**. The **residuals** function requires **fitted** and **hatvalues**. Finally, **fitted** requires **coef**.

We have brushed over one point about the use of inheritance in S3 classes. There are many more methods for objects of class *lm* than we have touched upon here:

```
> length(methods(class = "lm"))

[1] 38
```

and R will now assume that any of these methods can be used for our objects, because our objects inherit from class *lm*. However, calling these functions on our objects might raise undecipherable errors, or worse, nonsensical results that look plausible. To complete the class definition we should write simple functions for each of these that either works as intended or delivers a polite rejection.

3.4.9   Example Redux

The example fit has been scattered throughout the preceding sections. We can compare the parameter estimates with those obtained by using the `lm` function. We would hope that the parameter estimates would be close, and we would expect that the standard errors would be higher for the `lm` output for the parameters in the linear predictor.

```
> print(coef(summary(lm(height.m ~ dbh.cm, ufc))), digits = 3)

            Estimate Std. Error t value Pr(>|t|)
(Intercept)   12.676     0.5641    22.5 2.76e-72
dbh.cm         0.313     0.0139    22.5 1.64e-72
```

The parameter estimates all compare well with the output from the previous section.

Just to extend the example, and demonstrate the flexibility of the simple tools that we have constructed, we will now use them to fit a cubic spline model of height against diameter. First, we call the *splines* package to provide access to the relevant splines function, the enticingly named `bs`.

```
> library(splines)
```

Then we use the `bs` function, which constructs a set of linear predictors that corresponds to a cubic b-spline, in the model specification.

```
> ufc.g.spline <- ml_g(height.m ~ bs(dbh.cm),  data = ufc)
```

Before we interpret the model fit or parameter estimates we should check the regression diagnostics. As before, this is a simple matter thanks to our use of S3 classes (Figure 3.3).

```
> plot(ufc.g.spline)
```

The regression diagnostics compare very favorably with those provided in Figure 3.2. We can verify the improvement of the cubic spline model over the straight line model by comparing the log-likelihoods of the two models.

```
> logLik(ufc.g.reg)
```

```
'log Lik.' -1178.469 (df=3)
```

```
> logLik(ufc.g.spline)
```

```
'log Lik.' -1155.333 (df=5)
```

As a final step, we will write a function that performs an asymptotic likelihood ratio test using the reported log likelihoods of two models.

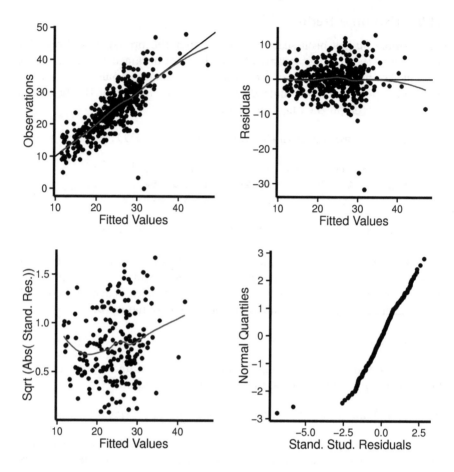

FIGURE 3.3
Four diagnostic graphs for the maximum-likelihood normal cubic spline regression of height against diameter using the tree measurement data from `ufc`.

```
> alrt <- function(x1, x2, ...) {
+    jll1 <- logLik(x1)
+    jll2 <- logLik(x2)
+    df1 <- attr(jll1, "df")
+    df2 <- attr(jll2, "df")
+    jll.diff <- abs(c(jll1) - c(jll2))
+    df.diff <- abs(df1 - df2)
+    p.value <- 1 - pchisq(2 * jll.diff, df = df.diff)
+    results <- list(out.tab = data.frame(model = c(1,2),
+                                          jll = c(jll1, jll2),
```

```
+                                              df = c(df1, df2)),
+                        jll.diff = jll.diff,
+                        df.diff = df.diff,
+                        p = p.value)
+      cat("\nLL of model 1: ", jll1, " df: ", df1,
+          "\nLL of model 2: ", jll2, " df: ", df2,
+          "\nDifference: ", jll.diff, " df: ", df.diff,
+          "\np-value against H_0: no difference between models ",
+          p.value, "\n")
+      return(invisible(results))
+ }
```

The code does not test whether the models are nested or whether the response variables are identical; the user is responsible for its sensible deployment.

Note that we have used the `invisible` function inside `return`. Recall that if the returned value of a function is not assigned to a name, then it is printed by default. Sometimes we prefer to control the format of output from within the function, but also to allow the user to be able to choose a name for the returned object and use it in other ways. The role of `invisible` is to prevent the object from being printed if it is not assigned to a name. Then, the function takes care of the printing of output, and the returned object is not printed.

We can now easily compare the fits of the two models using the asymptotic likelihood ratio test as follows.

```
> alrt(ufc.g.reg, ufc.g.spline)

LL of model 1:  -1178.469  df:  3
LL of model 2:  -1155.333  df:  5
Difference:  23.13595  df:  2
p-value against H_0: no difference between models  8.957479e-11
```

Here we see strong evidence of the improvement of fit. We could now ask whether simply adding a quadratic term would be sufficient improvement.

```
> ufc.g.quad <- ml_g(height.m ~ dbh.cm + I(dbh.cm^2),
+                    data = ufc)
> alrt(ufc.g.quad, ufc.g.spline)

LL of model 1:  -1164.706  df:  4
LL of model 2:  -1155.333  df:  5
Difference:  9.373229  df:  1
p-value against H_0: no difference between models  1.493004e-05
```

The likelihood ratio test suggests that the spline model provides a substantially better fit to the data than does the quadratic model.

3.4.10   Follow-up

Having fitted a suitable model, and studiously checked the graphical diagnostics, we are now positioned to use the model. We remind the reader of the three uses to which linear models are commonly put:

1. to enable prediction of a random variable at specific combinations of other variables;

2. to estimate the effect of one or more variables upon a random variable; and

3. to nominate a subset of variables that is most influential upon a random variable.

We have written methods that allow the resolution of each of these goals, in one way or another: `predict`, `coef`, and `alrt`, respectively. The first two of these functions require the same assumptions as for least-squares regression; the third requires that the errors be normally distributed, or the invocation of the Central Limit Theorem. We do not propose that these functions improve on the provided functions of R in any way, but we develop them to provide the reader insight into the linear model, and to programming in R.

3.5   Conclusion

In this chapter, we have blended the statistical ideas presented in Chapter 2 and the computing principles laid out in Chapter 1. We have introduced the approaches and code needed to fit linear models using least-squares and maximum likelihood regression. In our next chapter we will continue the maximum likelihood trajectory, allowing for conditional distributions from the exponential family: generalized linear models.

Many references are available for further reading about the theory and applications of least-squares linear regression. We particularly mention Draper and Smith (1998), Harrell (2001), Weisberg (2005), and especially Wood (2006).

3.6   Exercises

1. Given the definitions of matrices y and X below, where X is divided into two variables, [,1] and [,2], model y on X using matrices, determining the coefficients for the intercept and both X variables.

```
> y <- matrix(c(10,9,7,4,8,12,11,7,3,5,3,12,9,10), ncol=1)
> X <- matrix(c(1,4,6,0,5,4,1,4,7,1,2,2,1,4,
+                6,0,7,5,1,5,1,1,2,5,0,3,4,4),ncol=2)
```

2. Residuals:

   (a) What is the value of squaring the standardized deviance residuals from binary logistic models?

   (b) Why are deviance residuals preferred to Pearson residuals when assessing the fit of GLM models?

   (c) How may hat matrix diagonal statistics for a model be calculated using the standard error of the fitted value, μ?

   (d) Write a function for likelihood residual that is used with binomial models. You may have to do some research to find the definition; e.g., web search or Hilbe (2009).

4

Generalized Linear Models

4.1 Introduction

We have thus far discussed some basic types of statistical model estimation, providing code at the end of Chapter 3 to estimate the parameters of traditional linear regression models. These models required the assumption that the observed data were conditionally normal; that is, conditional on the model. The response variable followed the normal distribution. We now extend the linear model to include estimation of parameters of a collection of models that allow a range of conditional distributions — for example, binomial, Poisson, and gamma. These models are referred to as *generalized linear models*, or GLM. The linear model, being based on the normal probability distribution function (PDF) is a member of the GLM family. However, the normal model is nearly always estimated using the methods described in the last chapter, and we shall not cover it in this chapter.

We provide a rather thorough evaluation of GLM theory, families, and code due to the central role that GLMs now have in contemporary statistics. GLMs in various forms underlie ordered and unordered categorical response models, fixed, random and mixed effects models, hierarchical models, and a wide range of mixture models. More recently GLMs have begun to play an important role in understanding GLM-based Bayesian analysis. It is therefore wise to spend time on the fundamentals.

GLMs are all derived from the one-parameter exponential family of distributions. GLMs include discrete and continuous distributions, and can be modeled using standard maximum likelihood methods. In fact, most GLM models were estimated using maximum likelihood before a formal GLM methodology was developed. Moreover, with the exception of the normal model, GLM models that were based on continuous distributions were typically estimated as two-parameter models, specifically with parameters corresponding to location and scale. However, the formal GLM method does not directly estimate the scale parameter, but rather focuses on estimating the mean or location parameter. As we shall observe, though, depending on the needs of the researcher, relatively little modelling power is lost as a result. Note that the Gaussian or normal model also has a scale parameter, σ^2, which has until recently only rarely been estimated using maximum likelihood. We address two-parameter GLM models in the following chapter.

Generalized linear modelling as a methodology was developed in 1972 by John Nelder and Robert Wedderburn while working at the Rothamsted Experimental Station in the UK. Two years later, with a select group of statisticians associated with the Royal Statistical Society, Nelder and Wedderburn developed the Generalized Linear Interactive Models (G.L.I.M.) software application, which remained as the foremost GLM modelling tool for two decades. Peter McCullagh, of the University of Chicago, and Nelder authored the seminal text on the subject in 1983, with a second edition in 1989 (McCullagh and Nelder, 1989). Numerous books have subsequently been published on GLMs, as well as texts that extend GLM methodology to the modelling of longitudinal and clustered data, to GLM models incorporating smooth functions, and more recently to the implementation of the Bayesian analysis of GLM models. We believe that a solid case can be made that the essential GLM framework underlies much of the statistical developments of the final quarter of the 20th century.

A central feature of traditional GLM methodology is the iteratively reweighted least squares (IRLS) algorithm, which both linearizes the relationship between the model linear predictor and the fitted value, and provides a simple yet robust way to estimate model parameters. For this set of models, using the IRLS fitting algorithm simplifies the estimation process. See Hilbe (2011) for a history and derivation of IRLS methodology. IRLS is a simplification of the maximum likelihood algorithm that we discussed in the last chapter.

Prior to the development of GLM and GLM software, models currently estimated under the GLM framework were generally estimated using some variety of Newton–Raphson maximum likelihood estimation. We addressed this type of estimation in the last chapter, and will discuss it in considerable detail in Chapter 5 with respect to nonlinear models in general. For now it is important to keep in mind that IRLS yields maximum likelihood estimates, but is a simplified algorithm that allows easier and usually faster estimation of parameters.

In this chapter we shall provide an overview of the IRLS method of estimation, and will develop IRLS software in R that estimates parameters, standard errors, confidence intervals, and other ancillary statistics that are traditionally provided in GLM model output. We shall initially focus on the estimation of a single model, and then expand to provide a modular algorithm that incorporates all of the traditional GLM families, except the normal or Gaussian. The resultant function we develop, called `irls`, will appear somewhat similar to R's `glm` function; however, we shall structure the binomial response and parameterize the negative binomial heterogeneity parameter in a different manner to that of `glm`. They shall be framed in a manner similar to that employed in Stata's *glm* command, SAS's *Genmod* procedure, SPSS's *Genlin* procedure, and in the other GLM facilities residing in current commercial statistical software. There is a very good reason to do this, as shall be discussed later in the chapter.

4.2 GLM: Families and Terms

Generalized linear modelling is a method by which the standard linear regression model is extended to allow estimation of a certain set of traditional non-linear models. In particular, regression models based on the exponential family of distributions can be formulated as members of the family of generalized linear models if they meet the following criteria:

1. The response term, y, is conditionally distributed according to a member of the single parameter exponential family of probability distributions.

2. A monotonic and differentiable link function exists that linearizes the relationship between the linear predictor, η, or $\mathbf{X}\beta$, and the model fit, μ. This link function is typically symbolized as $g(\mu)$.

3. An inverse link function, $g^{-1}(\eta)$, exists that defines the model fit term.

4. Except for the Gaussian family, the variance, $V(\mu)$, is a function of the mean, which is the estimated model fit, μ. In the case of Gaussian-based models, the variance is fixed at 1.

5. GLMs are traditionally estimated using an Iteratively Re-weighted Least Squares algorithm, or IRLS, with a convergence criterion based on the change in either the deviance or log-likelihood function.

GLM families, and their associated variance functions, are traditionally based on the probability distributions given in Table 4.1.

TABLE 4.1
GLM distributions and variance functions.

PDF	Family	Variance
Continuous	Gaussian	1
	Gamma	μ^2
	Inverse Gaussian	μ^3
Discrete	Bernoulli	$\mu(1-\mu) = \mu - \mu^2$
	Binomial	$\mu(1-\mu/m) = \mu(m-\mu)/m$
	Poisson	μ
	Geometric	$\mu + \mu^2 = \mu(1+\mu)$
	Negative Binomial	$\mu + \alpha\mu^2 = \mu(1+\alpha\mu)$

Note that the Bernoulli distribution is the basis of binary logistic regression, and the binomial distribution — with m as the binomial denominator

— is the foundation of grouped logistic regression. When $m = 1$ for all observations in a model, the grouped logistic model is the same as the traditional binary response logistic regression. Authors of GLM software typically merge the Bernoulli family or model into the binomial. The default denominator is 1, which provides for a binary logistic model, but users may also declare alternative values for m, which generates a grouped or proportional model.

Note also that the geometric family is a subset of the negative binomial, with the heterogeneity parameter, which is commonly referred to as *alpha* (α), equal to 1. The Poisson may also be considered as a subset of the negative binomial, with $\alpha = 0$. This subject is considered in detail in Hilbe (2011). It is also interesting to observe that the Bernoulli and geometric variance functions only differ by a sign between the mean and mean-squared terms. The Bernoulli variance is rarely displayed as shown to the right in Table 4.1.

Table 4.2 displays the standard GLM families and their associated deviance and log-likelihood functions. When used for GLM estimation, both the deviance and log-likelihood are typically parameterized in terms of μ. When GLM models are estimated separately using a full maximum likelihood algorithm, the log-likelihood is always employed in place of the deviance. It is parameterized in terms of $\mathbf{X}\beta$. Most GLM algorithms use the deviance as the basis of convergence and for goodness-of-fit tests, with the log-likelihood being calculated after convergence. In any case, the deviance is defined in terms of the log-likelihood, $\mathcal{L}$, as

$$D = 2\sum_{i=1}^{n}\{\mathcal{L}\left(y_i; y_i\right) - \mathcal{L}(\mu_i; y_i)\} \qquad (4.1)$$

with the first term of the function indicating that every instance of the model fit, μ, is given the value of the response, y, and the second term of the function indicating the model log-likelihood function. Regardless of its definition, though, the deviance is generally an easier function to program and use in the GLM algorithm than its log-likelihood counterpart, and for that reason has retained its popularity.

The three continuous distributions provided here – namely Gaussian or normal, gamma, and inverse Gamma – are often estimated as two-parameter models, with the primary parameter being the model fit, μ, and the secondary parameter being the scale. GLM fits do not provide an estimate of the scale parameter directly, although for well-fitted models the scale may generally be approximated by the Pearson dispersion statistic. Note that the deviance functions that are used as objective functions for parameter estimation do not include the scale. Log-likelihoods are at times used in the IRLS algorithm in place of the deviance for discrete member models, but for the Gaussian, Inverse Gaussian, and gamma algorithms, the deviance is used to avoid the need of estimating a scale parameter. We shall henceforth refer only to the deviance in our discussion since it is the traditional basis of convergence. Be aware, however, that some implementations employ the log-likelihood.

TABLE 4.2

GLM deviance and log-likelihood functions for members of the exponential family(μ; $\mathbf{X}\beta$). x_+ indicates $\max(x, 1)$.

Gaussian	$\sum(y - \mu)^2$
LL(glm)	$-0.5 \sum\{(y - \mu)^2 + \log(2\pi)\}$
LL(μ)	$\sum\{(y \times \mu - \mu^2/2)/\sigma^2 - y^2/2\sigma^2 - 0.5 \times \log(2\pi\sigma^2)\}$
LL($\mathbf{X}\beta$)	$\sum\{[y \times (\mathbf{X}\beta) - (\mathbf{X}\beta)^2/2]/\sigma^2 - y^2/2\sigma^2 - 0.5 \times \log(2\pi\sigma^2)\}$
Bernoulli	$2\sum\{y \times \log(1/\mu) + (1 - y) \times \log(1/(1 - \mu))\}$
	$[2\sum\{y \times \log(y/\mu) + (1 - y) \times \log((1 - y)/(1 - \mu)\}]$
LL(μ)	$\sum\{y \times \log(\mu/(1 - \mu)) + \log(1 - \mu)\}$
LL($\mathbf{X}\beta$)	$\sum\{y \times (\mathbf{X}\beta) - \log(1 + exp(\mathbf{X}\beta))\}$
Binomial	$2\sum\{y \times \log(1/\mu) + (m - y) \times \log(1/(m - \mu))\}$
	$[2\sum\{y \times \log(y_+/\mu) + (m - y) \times \log((m - y)_+/(m - \mu)\}]$
LL(μ)	$\sum\{y \times \log(\mu/m) + (m - y) \times \log(1 - \mu/m) + \log\Gamma(m + 1) - \log\Gamma(y + 1) + \log\Gamma(m - y + 1)\}$
LL($\mathbf{X}\beta$)	$\sum\{y \times \log(\exp(\mathbf{X}\beta)/(1 + exp(\mathbf{X}\beta))) - (m - y) \times \log(\exp(\mathbf{X}\beta) + 1) + \log\Gamma(m + 1) - log\Gamma(y + 1) + \log\Gamma(m - y + 1)\}$
Poisson	$2\sum\{y \times \log(y/\mu) - (y - \mu)\}$
LL(μ)	$\sum\{y \times \log(\mu) - \mu - \log\Gamma(y + 1)\}$
LL($\mathbf{X}\beta$)	$\sum\{y \times (\mathbf{X}\beta) - \exp(\mathbf{X}\beta) - \log\Gamma(y + 1)\}$
NB2	$2\sum\{y \times \log(y/\mu) - (y + 1/\alpha) \times \log((1 + \alpha y)/(1 + \alpha\mu))\}$
LL(μ)	$\sum\{y \times \log((\alpha\mu)/(1 + \alpha\mu)) - (1/\alpha) \times \log(1 + \alpha\mu) + \log\Gamma(y + 1/\alpha) - \log\Gamma(y + 1) - \log\Gamma(1/\alpha)\}$
LL($\mathbf{X}\beta$)	$\sum\{y \times \log(\alpha \times \exp(x\beta)/(1 + \alpha \times \exp(x\beta))) - log(1 + \alpha \times \exp(x\beta))/\alpha + \log\Gamma(y + 1/\alpha) - \log\Gamma(y + 1) - \log\Gamma(1/\alpha)\}$
NBC	$\sum\{y \times (x\beta) + (1/\alpha) \times \log(1 - \exp(x\beta)) + \log\Gamma(y + 1/\alpha) - \log\Gamma(y + 1) - \log\Gamma(1/\alpha)\}$
Gamma	$2\sum\{(y - \mu)/\mu - \log(y/\mu)\}$
LL(glm)	$2\sum\{log(1/\mu) - (y/\mu)\}$
LL(μ)	$\sum\{((y/\mu) + \log(\mu))/- \phi + \log(y) \times (1 - \phi)/\phi - \log(\phi)/\phi - \log\Gamma(1/\phi)\}$
LL($\mathbf{X}\beta$)	$\sum\{(y \times (\mathbf{X}\beta) - \log(\mathbf{X}\beta))/- \phi + \log(y) \times (1 - \phi)/\phi - \log(\phi)/\phi - \log\Gamma(1/\phi)\}$
Inv Gaus	$\sum\{(y - \mu)^2/(y \times \mu^2)\}$
LL(glm)	$-1/2\sum\{(y - \mu)^2/(y\mu^2) + 3 \times \log(y) + \log(2\pi)\}$
LL(μ)	$\sum\{[(y/(2\mu^2)) - 1/\mu]/- \sigma^2 + 1/(-2y\sigma^2) - 0.5 \times log(2\pi y^3\sigma^2)\}$
LL($\mathbf{X}\beta$)	$\sum\{y/(2\mathbf{X}\beta) - sqrt(\mathbf{X}\beta)/- \sigma^2 + 1/(-2y\sigma^2) - 0.5 \times log(2\pi y^3\sigma^2)\}$

4.3   The Exponential Family

The PDF of the exponential family of distributions is typically expressed by
Equation 4.2.

$$f(y; \theta, \phi) \;=\; \exp\left\{ \frac{y_i \theta_i - b(\theta_i)}{\alpha_i(\phi)} \;+\; c(y_i; \phi) \right\} \tag{4.2}$$

where

θ_i is the canonical parameter or link function,

$b(\theta_i)$ is the cumulant,

$\alpha(\phi)$ is the scale parameter, set to one in discrete and count models, and

$C(y_i; \phi)$ is the normalization term, guaranteeing that the probability function
sums to unity.

The link and cumulant are the two foremost terms of interest of the exponential family. The link function, as mentioned before, is intended to linearize
the relationship between the linear predictor, $\mathbf{X}\beta$, and the model fit, μ. The
Greek letter η is the traditional GLM symbol for the inverse link. In the case
of the normal, or Gaussian, distribution, $\mu = \mathbf{X}\beta$. Since the two terms are
identical in this case, this particular link function is referred to as the *identity
link*. The standard formulations of the GLM link, inverse link, and derivative
of the link with respect to μ are displayed in Table 4.3 below.

TABLE 4.3
GLM link and inverse link functions.

Name	Link	Inverse Link	d(link)/dμ
Cloglog	$\log(-\log(m - \mu))$	$m - \exp(-e^{-\eta})$	$((\mu - m) \times \log(1 - \mu/m))^{-1}$
Identity	μ	η	1
Inverse	$1/\mu$	$1/\eta$	μ^{-2}
Logit	$\log(\mu/(m - \mu))$	$m/(1 + e^{-\eta})$	$m/(\mu(m - \mu))$
Probit	$\Phi^{-1}(\mu/m)$	$m\Phi(\eta)$	$(m * dnorm(qnorm(\mu/m)))^{-1}$
Log	$\log(\mu)$	e^{η}	$1/\mu$
NB-C	$-\log(1/(\alpha\mu) + 1)$	$1/(\alpha(e^{-\eta}) - 1))$	$1/(\mu + \alpha\mu^2)$
Sq. inv.	$1/\mu^2$	$1/1/\sqrt{\eta}$	$-2\mu^{-3}$

NB: the fitted value of the binary logistic model is π, which is the probability that the response, y, is 1. For the binary logistic model, the mean,
$\mu = \pi$. However, when the logistic model has a binomial denominator (m)
that is greater than 1, $\mu = m\pi$. The variance function is then $m\pi(1 - \pi)$, or
$\mu(1 - \mu/m)$. Note that $\mu(1 - \mu/m) = \mu(m - \mu)/m$.

The first derivative of the cumulant with respect to θ is the distributional mean; the second derivative is the variance. For example, the cumulant of the Bernoulli distribution is given as $-\log(1-\mu)$. The first derivative of this term, with respect to θ, which for the Bernoulli distribution is $\log(\mu/(1-\mu))$, is simply μ. μ, then, is the mean of the Bernoulli distribution. The second derivative is $\mu(1-\mu)$, the variance function. That is,

$$b'(\theta_i) = \text{mean}$$
$$b''(\theta_i) = \text{variance}$$

The canonical link function is derived directly from the PDF of a given GLM family. However, several well-known GLM models are often used with other link functions, called non-canonical link functions, e.g., the binomial distribution is often fit using the probit link or the complementary log–log link, and the negative binomial and gamma models are often fit using the log link. It is not clear that there is any statistical benefit in the canonization of a particular link function for each family. Different link functions lead to different interpretations of parameter estimates and also to different functional patterns, and should be selected based on the problem at hand, rather than on the basis of mathematical elegance. Theoretically, a GLM family can have an infinite number of different link functions. In fact, a general power link is included with many commercial GLM packages, thus providing a continuous range of power links. Generally, feasible values range from -3 to $+3$, but for some families/links, feasible values are within a more narrow range of power values. Commonly used powers are displayed in Table 4.4. Several are identical to the canonical link for the given family. The utility of power links rests with the fact that any intermediate value of a power can be used for a model. For example, based on comparative deviance, AIC and BIC statistics, it may be the case that a gamma model with a power link value of -1.225 is the best-fitted model. However, interpreting the model that is fitted with such a value may be quite another matter altogether.

TABLE 4.4
Some familiar members of the power link family.

Power	Name	Model Canonical Link
3	cube	none
2	square	none
1	identity	Gaussian
0	log	Poisson
0.5	square root	none
-1	inverse	gamma
-2	inverse quadratic	inverse Gaussian
-3	inverse cubic	none

One of the attractive features of GLMs is that it is easy to convert between a wide variety of models. One may change families by exchanging the deviance and associated weight functions. If the links also differ, then the link and inverse link may be exchanged as well. A canonical Poisson model can be made into a log-gamma model by changing only the deviance and weight functions. All else in the IRLS estimating algorithm remains the same.

When modelling within a family, e.g., the Bernoulli, one needs only change the link, inverse link, and weight in the IRLS estimating algorithm to change from a logistic to a probit model. That is, a binary probit regression differs from the logistic due only to the differing link and weight functions. The same is the case for a complementary log-log model. We will later capitalize on this convenience.

4.4   The IRLS Fitting Algorithm

The Iteratively Re-Weighted Least Squares (IRLS) algorithm may take several forms. Employing the traditional symbols for the link function, $g(\mu)$, the inverse link, $g^{-1}(\eta)$, which defines μ, and W as the weight, the IRLS algorithm may be expressed by the following schema (subscripts not displayed):

1. Initialize the expected response, μ, and the link function, $g(\mu)$. η is initialized as equal to $g(\mu)$.

2. Compute the weights as

$$W^{-1} = V g'(\mu)^2 = V \left(\frac{d\eta}{d\mu} \right)^2 \qquad (4.3)$$

where $g'(\mu)$ is the derivative of the link function and V is the variance, defined as the second derivative of the cumulant, $b''(\theta)$. For canonical links the derivative of the link is the inverse of the variance, resulting in $W = V$.

3. Compute a working response, which is a one-term Taylor linearization of the log-likelihood function, with a standard form of

$$z = \eta + (y - \mu) g'(\mu) \qquad (4.4)$$

4. Regress z on predictors $X_1 \ldots X_n$ with weights, W, to obtain updates on the vector of parameter estimates, β.

$$\beta_r = (X'WX)^{-1} X'Wz \qquad (4.5)$$

5. Compute η, or $X\beta$, the linear predictor, based on the regression estimates.

6. Compute μ, or $E(y)$, as $g^{-1}(\eta)$.

7. Compute the deviance function.

8. Iterate until the change in deviance between two iterations is below a specified level of tolerance.

Again, there are many possible modifications to the above scheme. However, most traditional GLM software implementations use methods similar to the above.

The GLM IRLS algorithm for the general case is presented in Table 4.5. The algorithm can be used for any member of the GLM family. Again, the substitution of specific functions into the general form for link, $g(\mu)$, inverse link, $g^{-1}(\eta)$, weight, W, and deviance or log-likelihood functions create different GLM models. All other aspects of the algorithm remain the same, hence allowing the user to easily change models.

Typically, with parameter estimates being of equal significance, the preferred model is the one with the lowest deviance as well as the lowest AIC or BIC statistic. AIC is the acronym for *Akaike Information Criterion* (Akaike, 1973), and BIC the acronym for *Bayesian Information Criterion*, both of which are based on the log-likelihood. The first form of BIC is due to the work of Schwarz (1978), who employed the log-likelihood in its equation. One form of BIC is based on the deviance (Raftery, 1986) and was commonly used in GLM software until developers began to provide the log-likelihood statistic in model output. Fit statistics for GLM models are described later in this chapter. Since nearly all of the post-estimation fit statistics use the model log-likelihood function, it is always important to calculate its value, which is typically accomplished following model convergence.

4.5 Bernoulli or Binary Logistic Regression

We can develop a schematic IRLS logistic regression algorithm given the values from the preceding tables. The terms in Table 4.6 define a binary logistic GLM algorithm. The weight is identical to the variance function for the family, and is inversely proportional to the derivative of the link with respect to μ. Since the logit link is canonical or natural to the binomial distribution, and therefore to the Bernoulli as well, the weight function for a logistic model is identical to its variance. This identity is the case for all canonically linked models.

The traditional form of the Bernoulli deviance is displayed in brackets in Table 4.2. Naive computation of the traditional expression of the binomial deviance function results in an error when $y = 0$. This is because the deviance requires the computing of $0 \times \log(0)$, which computers cannot handle without specific programming, because $log(0) = -\infty$. Using l'Hôpital's rule,

TABLE 4.5
Standard GLM IRLS estimating algorithm; p is the number of model predictors including the constant and n is the number of observations in the dataset. The algorithm allows for an offset.

dev $\leftarrow 0$	
ΔDev $\leftarrow 2 \times$ tolerance	
$\mu \leftarrow$ mean(y))	# initialize
$\eta \leftarrow g(\mu)$	# initialize link
while (abs(ΔDev) > tolerance){	# start loop; define convergence
$\quad w \leftarrow 1/(V \times g'^2)$	# weight
$\quad z \leftarrow \eta + (y - \mu)g' -$ offset	# working response
$\quad \beta = (X'wX)^{-1}X'wz$	# weighted regression
$\quad \eta \leftarrow \mathbf{X}\beta +$ offset	# linear predictor
$\quad \mu \leftarrow g^{-1}(\eta)$	# fitted value
$\quad$Dev0 $\leftarrow$ Dev	# copy deviance to Dev0
$\quad$Dev $\leftarrow$ Deviance function	
$\quad \Delta$Dev $\leftarrow$ Dev $-$ Dev0	# difference betw. new and old values
}	

TABLE 4.6
Functions required for using IRLS to fit a binary logistic GLM.

Link	$\log(\mu/(1-\mu))$
Inverse link	$1/(1+\exp(-\eta))$
Weight	$1/(\mu \times (1-\mu)) \times (1/(\mu \times (1-\mu)))^2 = \mu \times (1-\mu)$
Deviance	$2\Sigma(y \times \log(1/\mu) + (1-y) \times \log(1/(1-\mu)))$

$$\lim_{x \to 0+} x \times \log(x) = 0 \tag{4.6}$$

We have therefore restructured the general binomial deviance function to be

$$2\sum(y \times \log(1/\mu) + (1-y) \times \log(1/(1-\mu))) \tag{4.7}$$

Programmers will also sometimes partition the Bernoulli, or more general binomial deviance function, into separate expressions for the deviance for $y = 0$ and $y > 0$.

A simple working R script for a binary logistic regression using data from the 1912 Titanic ship disaster is shown below. Instead of convergence being based on the difference in deviance values between two iterations, we simply iterate four times. Most decently fitted logistic models usually need only three to five iterations before finding the appropriate parameter estimates. Note that the algorithm is specific to the `titanic` data. The data comprises a count response of the number of passengers who survived the accident, survived, and three predictors with the following format:

age: 1 = adult; 0 = youth
sex: 1 = male; 0 = female
class: 1 = first; 2 = second; 3 = third

We shall use only a single binary predictor, age, for this script

```
> library(msme)
> data(titanic)

> y <- titanic$survived
> x <- titanic$age
> mu <- rep(mean(y), nrow(titanic))    # initialize mu
> eta <- log(mu/(1-mu))                # initialize eta
> for (i in 1:4) {                     # loop for 4 iterations
+     w <-  mu*(1-mu)                   # weight = variance
+     z <- eta + (y - mu)/(mu*(1-mu))   # working response
+     mod <- lm(z ~ x, weights = w)     # weighted regression
+     eta <- mod$fit                    # linear predictor
+     mu <- 1/(1+exp(-eta))             # fitted value
+     cat(i, coef(mod), "\n")           # display iteration log
+ }

1 0.117651 -0.6658327
2 0.09178935 -0.6403553
3 0.09180755 -0.6403735
4 0.09180755 -0.6403735
```

The cat function results in the iterated output. The parameter estimates can then be found directly from the fitted linear model. The model coefficients may be abstracted from the above model, which we named mod, by issuing the following code. Recall that the variable x is identical to *age* in the titanic data.

```
> coef(summary(mod))
```

	Estimate	Std. Error	t value	Pr(>|t|)
(Intercept)	0.09180755	0.1919130	0.4783811	0.632458588
x	-0.64037350	0.2010116	-3.1857538	0.001477722

and 95% confidence intervals may be calculated as:

```
> confint(mod)
```

	2.5 %	97.5 %
(Intercept)	-0.2846818	0.4682969
x	-1.0347123	-0.2460347

We now compare our results those of R's `glm` function, which produces the following output.

```
> glm.test <- glm(survived ~ age,
+                      family = binomial,
+                      data = titanic)
> coef(summary(glm.test))
```

```
                 Estimate Std. Error   z value    Pr(>|z|)
(Intercept)   0.09180755  0.1917671  0.478745 0.632120053
age          -0.64037350  0.2008588 -3.188177 0.001431727
```

The coefficients of `mod` and `glm.test` are the same, but the standard errors differ slightly. It is important to know why this is the case. The standard errors from our model fit are computed from the last-fitted lm object, called `mod`. Because it is a linear model, the standard errors are scaled by the estimated standard deviation of the residuals, as we see from the code below.

```
> getAnywhere(vcov.lm)
```

```
A single object matching 'vcov.lm' was found
It was found in the following places
  registered S3 method for vcov from namespace stats
  namespace:stats
with value
```

```
function (object, ...)
{
    so <- summary.lm(object)
    so$sigma^2 * so$cov.unscaled
}
<bytecode: 0x18da684>
<environment: namespace:stats>
```

In order to obtain standard error estimates from an unscaled covariance matrix, we will need to use the following code.

```
> (se <- sqrt(diag(summary(mod)$cov.unscaled)))
```

```
(Intercept)           x
  0.1917671    0.2008588
```

These results are the same as reported by `glm`. We now make a brief detour to explore this difference in greater detail.

The `vcov.glm` function in R reports the model standard errors that are adjusted by the square root of the Pearson dispersion statistic, conditional on the model. For example, if the model is binomial, then the dispersion is 1, by

definition. If the model is quasi-binomial, which allows for non-unity dispersion, then the reported dispersion is computed from the model and the data.
The use of this dispersion to inflate or deflate the estimated standard errors is
called *scaling*. The Pearson dispersion is the ratio of the model Pearson Chi2
to the model degrees of freedom, which is defined as the number of observations in the model less the number of predictors, including the intercept. The
code for determining the dispersion is given later.

Standard errors are commonly scaled when there is evidence of over- or
under-dispersion in the data, such as might be caused by an inappropriate
variance function or unmodeled correlation. Correlation can arise from a variety of sources, e.g., when the observations have hierarchical structure, or
when an important term is missing from the model. In any case, when the
vcov function is used to determine standard errors, they are scaled, using the
dispersion that is dictated by the model.

If non-unity dispersion seems possible, then the glm function should be
used with the quasibinomial family. This is simply the binomial (logit) model
with standard errors multiplied by the square root of the Pearson dispersion
statistic as estimated from the model and data. Following our discussion of
scaling below, we address its further implications in Section 4.10.

```
> glm.qb <- glm(survived ~ age,
+               family = quasibinomial,
+               data = titanic)
> summary(glm.qb)$dispersion

[1] 1.001522

> coef(summary(glm.qb))

              Estimate Std. Error   t value     Pr(>|t|)
(Intercept)  0.09180755  0.1919130  0.478381 0.632458602
age         -0.64037350  0.2010116 -3.185754 0.001477722
```

We display (truncated) Stata statistical software output of the same model
using scaled standard errors. Compare the standard errors with the above
quasibinomial model.

```
. glm survived age, fam(bin) scale(x2) nolog nohead

------------------------------------------------------
             |                 OIM
    survived |    Coef.   Std. Err.      z    ...
-------------+----------------------------------------
         age | -.6403735   .2010116   -3.19   ...
       _cons |  .0918075   .191913     0.48   ...
------------------------------------------------------
(Standard errors scaled using square root of Pearson X2-based disp.)
```

Finally, we turn to the estimated confidence intervals. The confidence intervals from our bespoke fitted model are based on the scaled standard errors, so are not reliable in this instance.

```
> confint(mod)

                2.5 %       97.5 %
(Intercept) -0.2846818   0.4682969
x           -1.0347123  -0.2460347
```

We compare these with the confidence intervals produced by `confint.glm`. Note that the confidence intervals produced by `confint.glm` are not Wald confidence intervals. That is, they are not calculated as $\hat{\beta} \pm 1.96 \times \hat{s}_\beta$.

```
> confint(glm.test)

                2.5 %       97.5 %
(Intercept) -0.2840767   0.4698629
age         -1.0360643  -0.2467708
```

Notice the message directly under the `confint(glm.text)` function: `"Waiting for profiling to be done..."`. This refers to the fact that R is performing a profile likelihood of the coefficients over a range of coefficient values. The likelihood ratio test is used as the basis of the profiling. Traditional model-based 95% Wald confidence intervals may be obtained using the `confint.default` function, although these intervals are known to have poorer qualities, so profile-based intervals are preferred (see e.g., Pawitan, 2001).

```
> confint.default(glm.test)

                2.5 %       97.5 %
(Intercept) -0.2840491   0.4676642
age         -1.0340495  -0.2466975
```

These values may again be compared to using Stata. The results are identical to the standard errors and confidence intervals we calculated for the Stata scaled model and for the standard logistic model displayed below,

```
. glm survive age, fam(bin) nolog nohead

-----------------------------------------------------------------------
             |                 OIM
    survived |     Coef.   Std. Err.      [95% Conf. Interval]
-------------+--------------------- ... ---------------------------------
         age | -.6403735   .2008588  ...   -1.03405    -.2466975
       _cons |  .0918075   .1917671  ...   -.2840491    .4676642
-----------------------------------------------------------------------
```

Let us return to the `mod` results and create 95% confidence intervals by hand using the adjustment we made to have model-based rather than scaled standard errors. Given the rounding effect by using 1.96 for a alpha = 0.05, the results are identical to `glm` using the `confint.default` function and to the Stata results.

```
> mod$coef - 1.96*se

(Intercept)          x
 -0.284056    -1.034057

> mod$coef + 1.96*se

(Intercept)          x
  0.4676711   -0.2466902
```

4.5.1   IRLS

A more sophisticated R function for GLM logistic regression is given next. Initialization code is used to define the model, the response, *y*, and the matrix of model predictors, **X**, as done in Chapter 3 for maximum likelihood linear regression. This same initialization code shall be used for all of our maximum likelihood functions. Recall that it allows factoring of categorical variables, interactions, and even splines (refer to Section 3.4.9). This block of code is a powerful tool for fitting a wide range of linear models.

```
> irls_logit <- function(formula, data, tol = 0.000001) {# arguments
+
+ ## Set up the model components
+     mf <- model.frame(formula, data)        # define model frame
+     y <- model.response(mf, "numeric")      # set model response
+     X <- model.matrix(formula, data = data) # predictors in X
+
+ ## Check for missing values; stop if any.
+     if (any(is.na(cbind(y, X)))) stop("Some data are missing.")
+
+ ## Initialize mu, eta, the deviance, etc.
+     mu <- rep(mean(y), length(y))
+     eta <- log(mu/(1-mu))
+     dev <- 2 * sum(y*log(1/mu) +
+             (1 - y) * log(1/(1-mu)))
+     deltad <- 1
+     i <- 1
+
+ ## Loop through the IRLS algorithm
+     while (abs(deltad) > tol ) {             # IRLS loop begin
+         w <- mu * (1-mu)                     # weight
+         z <- eta + (y - mu)/w                # working response
```

```
+          mod <- lm(z ~ X-1, weights = w)        # weighted regression
+          eta <- mod$fit                         # linear predictor
+          mu <- 1/(1+exp(-eta))                  # fitted value
+          dev.old <- dev
+          dev <- 2 * sum(y * log(1/mu) +
+                          (1 - y) * log(1/(1 - mu)))  # deviance
+          deltad <- dev - dev.old                # change
+          cat(i, coef(mod), deltad, "\n")        # iteration log
+          i <- i + 1                             # iterate
+     }
+
+ ## Build some post-estimation statistics
+     df.residual <- summary(mod)$df[2]
+     pearson.chi2 <- sum((y - mu)^2 / (mu * (1 - mu))) / df.residual
+     se.beta.hat <- sqrt(diag(summary(mod)$cov.unscaled))
+
+ ## Return a compact result
+     result <- list(coefficients = coef(mod),
+                      se.beta.hat = se.beta.hat)
+   return(result)
+ }
```

Note that the operations that are essential to a regression routine cannot be performed if any observation in the data matrix has a missing value. As before, we have made the user responsible for checking and perhaps correcting missing values, or deleting them from the data matrix. If we had wanted R to delete rows in which any missing value exists, we would add `data <- na.omit(data)` before we set up the model components.

The generic version of a logistic regression function given above uses the change in the deviance function as the basis of convergence. The standard criterion used with commercial software is 10^{-6} or 0.000001. Convergence is achieved when the absolute difference of two consecutive iterations of deviance function is less than 10^{-6}, which is one-millionth of unity. The two values of the deviance must therefore be nearly identical in order for convergence to be declared. Note that the residual degrees of freedom are created after the `while` loop, followed by calculating the Pearson Chi2 dispersion statistic. The square root of the dispersion is then used to modify `sqrt(diag(vcov(mod)))`.

We now use this function to fit a logit model to the *medpar* data.

```
> library(msme)
> data(medpar)
```

The `medpar` data is modeled using the `irls_logit` function as

```
> i.logit <- irls_logit(died ~ hmo + white,
+                        data = medpar)

1 -0.9129327 -0.01225269 0.2902311 -2.258749
2 -0.9261486 -0.01224646 0.3033496 -0.004542199
3 -0.9261862 -0.01224648 0.3033872 -3.655578e-08
```

The iteration log defined by i.logit reports the sequential convergence of the intercept, hmo, white, and the difference in deviance functions between two iterations. With the default tolerance set at 0.000001, or 10^{-6}, we note that convergence was achieved at iteration 4. Printing the name given to the model, i.logit, displays the vector of coefficients and of standard errors.

```
> i.logit
```

```
$coefficients
X(Intercept)          Xhmo          Xwhite
 -0.92618620   -0.01224648      0.30338724
```

```
$se.beta.hat
X(Intercept)          Xhmo          Xwhite
   0.1973889     0.1489250       0.2051781
```

95% Wald confidence intervals may easily be computed using the code:

```
> with(i.logit, coefficients - 1.96 * se.beta.hat)
```

```
X(Intercept)          Xhmo          Xwhite
  -1.3130684    -0.3041395      -0.0987619
```

```
> with(i.logit, coefficients + 1.96 * se.beta.hat)
```

```
X(Intercept)          Xhmo          Xwhite
  -0.5393040     0.2796466       0.7055364
```

The z-values and the corresponding p-values can be obtained by

```
> (Z <- with(i.logit, coefficients / se.beta.hat))
```

```
X(Intercept)          Xhmo          Xwhite
  -4.69219056   -0.08223254     1.47865289
```

```
> (pvalues <- 2*pnorm(abs(Z), lower.tail = FALSE))
```

```
X(Intercept)          Xhmo          Xwhite
2.702952e-06 9.344618e-01 1.392331e-01
```

A comparison of coefficients with glm can be obtained by

```
> glm.logit <- glm(died ~ hmo + white,
+                        family = binomial,
+                        data = medpar)
> coef(summary(glm.logit))
```

```
            Estimate Std. Error    z value      Pr(>|z|)
(Intercept) -0.92618620  0.1973903 -4.69215560 2.703414e-06
hmo         -0.01224648  0.1489251 -0.08223252 9.344618e-01
white        0.30338724  0.2051795  1.47864270 1.392358e-01
```

As before, values of the confidence intervals will differ slightly between glm and our stand-alone logistic model. Our code produces Wald confidence intervals. Recalling our earlier discussion, we may obtain Wald confidence intervals from *glm* objects by using the `confint.default` function

```
> confint.default(glm.logit)
```

```
               2.5 %     97.5 %
(Intercept) -1.31306417 -0.5393082
hmo         -0.30413424  0.2796413
white       -0.09875728  0.7055318
```

These confidence intervals are the same as produced using the `irls_logit` function. We reproduce them below with a different format.

```
> estim <-coef(summary(glm.logit))
> beta <- estim[,1]
> se <- estim[,2]
> confint.model <- (cbind(beta,
+                         beta - se*qnorm(.975),
+                         beta + se*qnorm(.975)))
> colnames(confint.model) <- c("Beta", "low_CI", "Hi_CI")
> confint.model
```

```
               Beta     low_CI      Hi_CI
(Intercept) -0.92618620 -1.31306417 -0.5393082
hmo         -0.01224648 -0.30413424  0.2796413
white        0.30338724 -0.09875728  0.7055318
```

The results are identical to those of `confint.default`.

4.6 Grouped Binomial Models

One of the nice features of the GLM algorithm is that it provides for the estimation of binomial or grouped logistic regression, as well as the binomial parameterizations of the non-canonical links. Recall that we have already been working with binomial models — models with a binomial denominator of 1. Programming and using binomial or grouped models requires the incorporation of a binomial denominator into the model. The denominator is the number of observations in the model sharing the same *covariate pattern*. The numerator is the number of successes (1's) for a given covariate pattern. Suppose we have observation data appearing as per Table 4.7.

TABLE 4.7
Observation-level binomial data.

Obs.	y	x1	x2	cp#
1:	1	0	0	1 (0,0)
2:	1	1	0	2 (1,0)
3:	0	0	1	3 (0,1)
4:	1	0	1	3
5:	1	1	1	4 (1,1)
6:	0	1	1	4
7:	0	0	0	1
8:	1	0	1	3
9:	1	1	0	2
10:	0	1	0	2
11:	0	0	1	3
12:	1	0	0	1
13:	0	1	1	4
14:	1	0	0	1

Each value of *cp#* represents a distinct covariate pattern, the value profile of the model predictors. A grouped dataset can be formed from the above by combining covariate patterns, but re-defining y as the number of 1's for each covariate pattern, and m the number of observations having the same covariate pattern. The grouped or proportional dataset is provided in Table 4.8.

TABLE 4.8
Grouped binomial data.

Obs.	y	m	x1	x2
1:	3	4	0	0
2:	2	3	1	0
3:	2	4	0	1
4:	1	3	1	1

Note that the sum of y values in the observation data equals the sum of y values in the grouped data. Likewise, the sum of m values equals the number of observations in the observation data. The data presented in both tables are identical; it is simply formatted in different ways. The grouped or binomial model is appropriate for data structured as Table 4.8. Data in tables can normally be converted to the grouped data format for analysis.

The `irls_logit` function we created to estimate the parameters of a binary logistic model can be amended to incorporate both an offset and a denominator greater than 1. The default value of *offset* is 0. Offsets are either added or subtracted to lines in the algorithm. By initializing the value as 0, if offsets are not used, then no change is made to the algorithm. The binomial denominator `m`, on the other hand, is assigned a default value of 1 since it serves as a multiplier to other terms in the model.

The formulae for the link, inverse link, weight, and deviance functions for binomial or grouped logistic GLM models are located in Tables 4.2 and 4.3. The deviance that is displayed in square brackets is the equation normally observed in texts on the subject. As noted earlier, we use a different version that does not require calculation of log(0).

Recall from Equation 4.3 that the GLM weight is

$$w^{-1} = V\left(\frac{dg\,(\mu)}{d\mu}\right)^2$$

where V is the GLM family variance function and $g(\mu) = g$ is the model link function. This equation can be simplified in appearance as

$$w^{-1} = Vg'^2 \tag{4.8}$$

For canonical links, $g' = 1/V$, so the terms cancel, leaving

$$w = g' = 1/V \tag{4.9}$$

Therefore, for the grouped binomial logit model, the values for the various components are provided in Table 4.9.

TABLE 4.9
Components of the grouped binomial logistic regression using IRLS.

Deviance	$2\sum y \times \log(1/\mu) + (m-y) \times \log(1/(m-\mu))$
Link function	$\log(\mu/(m-\mu))$
Inverse link	$m/(1 + \exp(-\eta))$
Logit weight	$\mu \times (m-\mu)/m$
Logit working response	$z = \eta + (y-\mu)/(\mu(1-\mu/m))$

We next provide a stand-alone function for the estimation of both binary and grouped logistic models. If `m` is not provided a value, then the algorithm assumes that y is binary $(0/1)$ and that the model is a binary logistic regression. If m is given a value, or assigned a variable, the algorithm assumes that a grouped logistic model is to be estimated. In all cases, the user must make certain that y is always equal to or less than m; i.e., $y \le m$. Finally, referring to Table 4.1, $\mu \times (1-\mu/m) = \mu \times (m-\mu)/m$. Both formulations are commonly used by programmers. Note also that we use the grouped binomial variance function, `mu * (1 - mu/m)`, in the calculation of the Pearson Chi2 statistic.

```
> irls_glogit <- function(formula, data,
+                          tol = 0.000001, offset = 0, m = 1 ) {
+
+ ## Set up the model components
+    mf <- model.frame(formula, data)
+    y <- model.response(mf, "numeric")
+    X <- model.matrix(formula, data = data)
+
+ ## Check for missing values; stop if any.
+    if (any(is.na(cbind(y, X)))) stop("Some data are missing.")
+
+ ## Check for nonsensical values; stop if any.
+    if (any(y < 0 | y > m)) stop("Some data are absurd.")
+
+ ## Initialize mu, eta, the deviance, etc.
+    mu <- rep(mean(y), length(y))
+    eta <- log(mu/(m - mu))
+    dev <- 2 * sum(y * log(pmax(y,1)/mu) +
+                (m - y)* log(pmax(y,1)/(m - mu)) )
+    deltad <- 1
+    i <- 1
+
+ ## Loop through the IRLS algorithm
+    while (abs(deltad) > tol ) {
+       w <- mu*(m-mu)/m
+       z <- eta + (y-mu)/w - offset
+       mod <- lm(z ~ X-1, weights=w)
+       eta <- mod$fit + offset
+       mu <- m/(1+exp(-eta))
+       dev.old <- dev
+       dev <- 2 * sum(y*log(pmax(y,1)/mu) +
+                   (m - y)* log(pmax(y,1)/(m - mu)))
+       deltad <- dev - dev.old
+       i <- i + 1
+    }
+
+ ## Build some post-estimation statistics
+    df.residual <- summary(mod)$df[2]
+    pearson.chi2 <- sum((y - mu)^2 /
+                   (mu * (1 - mu/m))) / df.residual
+
+ ## Return a brief result
+    return(list(coef = coef(mod),
+                se = sqrt(diag(summary(mod)$cov.unscaled)),
```

```
+                         pearson.chi2 = pearson.chi2))
+  }
```

The `pmax(y,1)` function gives the *parallel* maxima of the vectors named in its arguments.

```
> pmax(c(0, 1, 1), c(1, -1, 0))
```

```
[1] 1 1 1
```

For an example of a grouped logistic model we shall use the noted `doll` dataset displayed in Table 4.10 which was developed by Doll and Hill in 1966 (Doll and Hill, 1966). The data are a record of physician smoking habits and the probability of death by myocardial infarction. The physicians were divided into five age divisions, with `deaths` as the response, person years (`pyears`) as the binomial denominator, and both smoking behavior (`smoke`) and age group (*a1-a5*) as predictors. Here we will interpret `age` as an ordinal variable. The data can be located in `doll`, which is in the *msme* package.

TABLE 4.10
Doll data.

Age	Person years Non-Smokers	Smokers	Coronary deaths Non-Smokers	Smokers
35-44	18790	52407	2	32
45-54	10673	43248	12	104
55-64	5710	28612	28	206
65-74	2585	12663	28	186
75-84	1462	5317	31	102

```
> library(msme)
> data(doll)
```

```
> i.glog <- irls_glogit(deaths ~ smokes + ordered(age),
+                        data = doll,
+                        m = doll$pyears)
> i.glog
```

```
$coef
   X(Intercept)          Xsmokes Xordered(age).L
    -5.68086278       0.35780978      2.94655871
Xordered(age).Q Xordered(age).C Xordered(age)^4
    -0.71565635      -0.01185071      0.01474086
```

```
$se
```

```
X(Intercept)           Xsmokes Xordered(age).L
  0.10271860          0.10781097      0.12723208
Xordered(age).Q Xordered(age).C Xordered(age)^4
  0.11322768          0.09519349      0.07616658
```

```
$pearson.chi2
[1] 2.78052
```

Notice the substantial dispersion value. This clearly indicates either considerable correlation in the data, which violates the model assumption of the independence of observations, or under-fitting. Scaling is a commonly used method for adjusting standard errors in light of excessive correlation in the model data.

The parameter estimates for the model need some explanation. Because we declared **age** to be an ordinal variable, it has been reparameterized into a four-degree orthogonal polynomial, which requires the same number of degrees of freedom, but provides quite a different interpretation. Instead of estimating a mean for the first level and a set of four differences, which is the usual parameterization for (five-level) factors, this model guides the choice of a suitable polynomial by fitting increasing orthogonal powers: a constant, a linear model, quadratic, cubic, etc. Here, the z-values are

```
> with(i.glog, coef / se)
```

```
X(Intercept)           Xsmokes Xordered(age).L
 -55.3051024          3.3188624     23.1589294
Xordered(age).Q Xordered(age).C Xordered(age)^4
  -6.3205072         -0.1244908      0.1935345
```

which suggest that a quadratic model would be adequate for our purposes.

We check our results using R's `glm` function as follows.

```
> glm.glog <- glm(cbind(deaths, pyears - deaths) ~
+                 smokes + ordered(age),
+                 data = doll,
+                 family = binomial)
> coef(summary(glm.glog))
```

```
                Estimate Std. Error    z value
(Intercept)     -5.68086278 0.10271854 -55.3051331
smokes           0.35780978 0.10781093   3.3188636
ordered(age).L   2.94655871 0.12723192  23.1589577
ordered(age).Q  -0.71565635 0.11322755  -6.3205142
ordered(age).C  -0.01185071 0.09519343  -0.1244909
ordered(age)^4   0.01474086 0.07616657   0.1935345
                 Pr(>|z|)
(Intercept)     0.000000e+00
```

```
smokes              9.038456e-04
ordered(age).L  1.180987e-118
ordered(age).Q  2.606944e-10
ordered(age).C  9.009266e-01
ordered(age)^4  8.465404e-01
```

The coefficients and standard errors produced by `glm` and `irls_glogit` are the same.

In previous sections we demonstrated how to construct stand-alone GLM-type IRLS functions for the binary logistic and grouped logistic models. Next we put the various stand-alone GLM functions into a single algorithm.

4.7 Constructing a GLM Function

We now shall combine the above stand-alone GLM functions that we constructed in the previous sections, together with other GLM families, into a single function called `irls`. `irls` will allow the use of a `summary(<modelname>)` function after estimation, displaying a table of parameter estimates and associated standard errors, z-statistics, p-values, and 95% confidence intervals. Moreover, the `irls` function will provide users with a number of post-estimation statistics which can be used to assess fit, create graphs, or used to calculate other statistics. The user will also be able to access the call used to estimate model parameters and associated statistics by typing `modelname$call`, and may display the family used to model the data by typing `modelname$family`.

The key to constructing a multifunction function is modularity. That is, we gather equations defining the link, inverse link, and variance functions of the various GLM models into one module, the deviances and log-likelihoods in another module, and so forth. The name of these modules is defined and then applied in the function.

First, the `irls` function defines various generic functions that will be used within the GLM IRLS algorithm:

```
> jllm   <- function(y, mu, m, a) UseMethod("jllm")
> linkFn <- function(mu, m, a) UseMethod("linkFn")
> lPrime <- function(mu, m, a) UseMethod("lPrime")
> unlink <- function(y, eta, m, a) UseMethod("unlink")
> variance <- function(mu, m, a) UseMethod("variance")
```

The `lPrime` function is the first derivative of the link function with respect to its first argument. The 'm' in `jllm` signifies the mean parameterization.

We next define a collection of methods that will be used by the function for logit-linked binomial regression. The deviance is defined as the sum of squared deviance residuals. Software developers typically code the deviance and deviance residuals separately, using formulae like those presented in Table 4.2. Here we code the deviance residuals directly from the joint log-likelihood. The residuals are then squared and summed to produce the deviance statistic. This approach to coding does not produce speedy or robust execution. However, it does cleanly demonstrate the unity in the statistical ideas that underpin GLM. The deviance residuals are defined as

```
> devianceResids <- function(y, mu, m, a)
+    sign(y - mu) * sqrt(2 * abs(jllm(y, mu, m, a) -
+                             jllm(y, y,  m, a)))
```

and the deviance follows naturally,

```
> devIRLS <- function(object, ...)
+    sum(devianceResids(object, ...)^2)
```

We choose these peculiar names to distinguish these two functions from the more general ones that will be provided in the next chapter.

Note also that we have tried to anticipate all the possible parameters that we are likely to wish to pass to the `devianceResiduals` function. We are presently writing the `irls` function to cope with grouped and ungrouped logistic regression, so we need to include m, the group size. However, we will later wish to model the negative binomial distribution, which has a scale parameter a. To prevent us from having to redefine a more complete function later, we define it now.

An alternative would be to define these functions within the body of the `irls` function, which we will define shortly. Doing so would reduce the modularity of the code, so we elect to define all the needed arguments up front.

The parameter estimates for models of the exponential family are initialized using the following code.

```
> initialize <- function(y, m) {
+    ret.y <- rep(mean(y), length(y))
+    class(ret.y) <- class(y)
+    ret.y
+ }
```

Next we define the formulae for the variance and log-likelihood for the binomial family. Residuals are addressed further in Section 4.11.2.

```
> variance.binomial <- function(mu, m, a) mu * (1 - mu/m)
```

As noted in Section 1.2.3.3, we provide all the parameters that we are likely to need, as well as those that are needed now.

The joint log-likelihood is most efficiently provided by using R's internal function as discussed in Section 1.2.5.

```
> jllm.binomial <- function(y, mu, m, a)
+   dbinom(x = y, size = m, prob = mu / m, log = TRUE)
```

The next defined module relates to the link function. For each family, three functions are provided for the link, derivative of the link, and inverse link functions. For the binomial-logit link, the functions are

```
> linkFn.logit <- function(mu, m, a) log(mu / (m - mu))
> lPrime.logit <- function(mu, m, a) m / (mu * (m - mu))
> unlink.logit <- function(y, eta, m, a) m / (1 + exp(-eta))
```

Note that the `variance(mu)` function, the derivative of the link, `lPrime(mu)`, the inverse link, `unlink(eta)`, and deviance, `deviance(y,mu)`, functions have been pre-defined earlier in the algorithm. Also note that, as noted in Section 1.2.3.3, we have chosen a very specific strategy to communicate the family and the choice of link function to the algorithm: we use S3 classes (see Chapters 1 and 3). That is, we assign the family and the link function to the data as classes, and this information is carried through the algorithm automatically, and used by R to select the appropriate collection of functions. This is a hack that pivots on the flexible implementation of inheritance in S3 classes. Consequently, in the case of the `unlink` function, the sole purpose of including y among the arguments is to provide class information.

Hence, here all we do is call variance(mu), for example, and because `mu` has a specific class that is defined by `irls` to be the family name, R will know exactly which variance function to use. An alternative would have been to use multiple `if` statements, but the present approach provides cleaner code that is easier to read and maintain.

Next, the `irls` function is defined, and options specified together with their default values

```
> irls <- function(formula, data, family, link,
+                   tol = 1e-6,
+                   offset = 0,
+                   m = 1,
+                   a = 1,
+                   verbose = 0) {
+
+ ### Prepare the model components as previously
+   mf <- model.frame(formula, data)
+   y <- model.response(mf, "numeric")
+   X <- model.matrix(formula, data = data)
+
+ ### Check for missing values
+   if (any(is.na(cbind(y, X)))) stop("Some data are missing.")
+
+ ### Arrange the class information
```

```
+    class(y) <- c(family, link, "expFamily")
+
+ ### Establish a start point
+    mu <- initialize(y, m)
+    eta <- linkFn(mu, m, a)
+    dev <- devIRLS(y, mu, m, a)
+    deltad <- 1
+    i <- 0
+
+ ### IRLS loop (as before)
+    while (abs(deltad) > tol ) {
+      w <-  1 / (variance(mu, m, a) * lPrime(mu, m, a)^2)
+      z <- eta + (y - mu) * lPrime(mu, m, a) - offset
+      mod <- lm(z ~ X - 1, weights = w)
+      eta <- mod$fit + offset
+      mu <- unlink(y, eta, m, a)
+      dev.old <- dev
+      dev <- devIRLS(y, mu, m, a)
+      deltad <- dev - dev.old
+      i <- i + 1
+      if(verbose > 0) cat(i, coef(mod), deltad, "\n")
+    }
+
+ ### Post-estimation statistics
+    df.residual <- summary(mod)$df[2]
+    pearson.chi2 <- sum((y - mu)^2 /
+                    variance(mu, m, a)) / df.residual
+    ll <- sum(jllm(y, mu, m, a))
+    se.beta.hat <- sqrt(diag(summary(mod)$cov.unscaled))
+
+ ### Return a rich object --- allows use of print.glm
+    result <-
+      list(coefficients = coef(mod),
+           se.beta.hat = se.beta.hat,
+           model = mod,
+           call = match.call(),
+           nobs = length(y),
+           eta = eta,
+           mu = mu,
+           df.residual = df.residual,
+           df.null = length(y) - 1,
+           deviance = dev,
+           null.deviance = NA,
+           p.dispersion = pearson.chi2,
+           pearson = pearson.chi2 * df.residual,
```

```
+            loglik = ll,
+            family = list(family = family),
+            X = X,
+            i = i,
+            residuals = devianceResids(y, mu, m, a),
+            aic = -2 * ll + 2 * summary(mod)$df[1])
+    class(result) <- c("msme","glm")
+    return(result)
+ }
```

An optional iteration log is provided that displays the iteration number, i, the values of the model coefficients, `coef(mod)`, and the value of the difference in deviances for each iteration until convergence. `mod` is the name we assigned to the fitted weighted linear regression within the IRLS loop, which provides appropriate parameter estimates and standard errors.

Following convergence and the estimation of parameter estimates and fitted values, the `irls` algorithm calculates summary statistics based on the final values of `mu` and `eta`. These include the residual degrees of freedom, the Pearson `chi2` statistic, and log-likelihood. Thereafter, a result list is provided specifying which statistics are to be made available to the user as post-estimation statistics.

It is important to return a value for the null deviance, even though we do not compute it, because doing so enables us to use the `print.glm` function, which is very convenient. For `irls` we prefer to omit the computation of the null deviance, simply because adding the needed code would lengthen the function too much for our purposes. We show how it may be easily obtained shortly.

The results are given a specific class, as we showed in Chapter 3, in order that specially-written functions will be used when we print or plot the output object. All statistics defined in the `result` list are `return`ed for the use of those who are modelling the data.

Other statistics may also be defined from the returned statistics. `irls` creates several such statistics prior to defining a final module that displays the output given to the screen with the `summary` function.

We now demonstrate the new function using the `medpar` data and a model that we fitted earlier.

```
> irls.logit <- irls(died ~ hmo + white,
+                    family = "binomial", link = "logit",
+                    data = medpar)
> irls.logit

Call:  irls(formula = died ~ hmo + white, data = medpar,
    family = "binomial", link = "logit")

Coefficients:
```

```
X(Intercept)              Xhmo              Xwhite
    -0.92619          -0.01225           0.30339
```

```
Degrees of Freedom: 1494 Total (i.e., Null);  1492 Residual
Null Deviance:              NA
Residual Deviance: 1921            AIC: 1927
```

The reader can verify that these estimates are consistent with those obtained earlier. As mentioned above, our function does not compute the null deviance. The reader may calculate it by explicitly fitting an intercept-only model.

```
> with(irls(died ~ 1,
+            family = "binomial", link = "logit",
+            data = medpar), c(deviance, df.residual))
```

```
[1] 1922.865 1494.000
```

4.7.1 A Summary Function

We now have a function that will compute and report a logistic regression, given suitable inputs. We will write a summary method to complete our little collection. In early drafts of this project we did try to adopt the summary.glm method, which also required examining the print.summary.glm method, but we decided that satisfying the existing function looked like too much work[1], and that writing our own summary function would be more efficient.

```
> summary.msme <- function(object, ...) {
+
+ ### Create a coefficient table
+   z <- with(object, coefficients / se.beta.hat)
+   zTable <-
+       with(object,
+            data.frame(Estimate = coefficients,
+                       SE = se.beta.hat,
+                       Z = z,
+                       p = 2 * pnorm(-abs(z)),
+                       LCL = coefficients - 1.96*se.beta.hat,
+                       UCL = coefficients + 1.96*se.beta.hat))
+   rownames(zTable) <- colnames(object$X)
+
+ ### Prepare part of the coefficient table for printing
+   z.print <- zTable
+   z.print$p <- formatC(z.print$p, digits = 3, format="g")
```

---

[1] We invite the reader to consider it as an exercise, at no additional cost.

```
+
+ ### Build a list of output objects
+   summ <- list(call = object$call,
+                coefficients = zTable,
+                deviance = object$deviance,
+                null.deviance = object$null.deviance,
+                df.residual = object$df.residual,
+                df.null = object$df.null)
+
+ ### Write out a set of results
+   cat("\nCall:\n")
+   print(object$call)
+   cat("\nDeviance Residuals:\n")
+   print(summary(as.numeric(object$residuals)))
+   cat("\nCoefficients:\n")
+   print(z.print, digits = 3, ...)
+   cat("\nNull deviance:", summ$null.deviance,
+       " on ", summ$df.null, "d.f.")
+   cat("\nResidual deviance:", summ$deviance,
+       " on ", summ$df.residual, "d.f.")
+   cat("\nAIC: ", object$aic)
+   cat("\n\nNumber of optimizer iterations: ",
+       object$i, "\n\n")
+
+ ### Return the list but do not print it.
+   return(invisible(summ))
+ }
```

We can now compare the results of our summary function with the outcome of calling **summary** upon R's *glm*-classed object.

```
> summary(irls.logit)

Call:
irls(formula = died ~ hmo + white, data = medpar,
    family = "binomial", link = "logit")

Deviance Residuals:
   Min. 1st Qu.  Median     Mean 3rd Qu.     Max.
-0.9268 -0.9268 -0.9222 -0.1002  1.4510   1.5930

Coefficients:
              Estimate     SE       Z        p     LCL     UCL
(Intercept)   -0.9262  0.197  -4.6922  2.7e-06 -1.3131 -0.539
hmo           -0.0122  0.149  -0.0822    0.934 -0.3041  0.280
white          0.3034  0.205   1.4787    0.139 -0.0988  0.706
```

```
Null deviance: NA   on  1494 d.f.
Residual deviance: 1920.602  on  1492 d.f.
AIC:  1926.602


Number of optimizer iterations:  3


> summary(glm.logit)


Call:
glm(formula = died ~ hmo + white, family = binomial,
    data = medpar)


Deviance Residuals:
    Min       1Q    Median      3Q       Max
 -0.9268  -0.9268  -0.9222   1.4507    1.5929


Coefficients:
             Estimate Std. Error z value Pr(>|z|)
(Intercept) -0.92619    0.19739  -4.692  2.7e-06 ***
hmo         -0.01225    0.14893  -0.082    0.934
white        0.30339    0.20518   1.479    0.139
---
Signif. codes:  0 '***' 0.001 '**' 0.01 '*' 0.05 '.' 0.1 ' ' 1


(Dispersion parameter for binomial family taken to be 1)


    Null deviance: 1922.9  on 1494  degrees of freedom
Residual deviance: 1920.6  on 1492  degrees of freedom
AIC: 1926.6


Number of Fisher Scoring iterations: 4
```

The outputs of the summary functions for the results are very similar.

Thus far we have developed IRLS algorithms and functions that estimate binary and grouped logistic regression. A simple operation allows us to convert them to *probit* and *complementary log-log* (*cloglog*) models. It must be recalled, though, that *probit* and *cloglog* are non-canonical models. Their link functions are not derived from the binomial PDF. To fit a binary logistic model using a *probit* or *cloglog* link, the link, inverse link, and the weight functions must be amended. Recall that for canonical models the terms of the weight function cancel, leaving the inverse of the variance. But this is not the case for non-canonical weights. Since we constructed the `irls` functions in their unsimplified form, we merely need to provide new functions, as follows. Everything else about the algorithm remains the same.

4.7.2   Other Link Functions

We shall amend the `irls` function to allow fitting a probit model. Only the link, inverse link, and weight must be changed. Table 4.11 presents the other link functions, and the code for them follows immediately after.

TABLE 4.11
Probit/cloglog link functions, inverse links, and derivatives of links.

	probit	cloglog
Link function	$\Phi^{-1}(\mu)$	$\log(-\log(1-\mu))$
Inverse link	$\Phi(\eta)$	$1 - \exp(-\exp(-\eta))$
Derivative of link	$1/(\mathrm{dnorm}(\mathrm{qnorm}(\mu)))$	$1/((\mu-1) \times \log(1-\mu))$

```
> linkFn.probit <- function(mu, m, a) qnorm(mu / m)
> lPrime.probit <- function(mu, m, a)
+                         1 / (m * dnorm(qnorm(mu/m)))
> unlink.probit <- function(y, eta, m, a) m * pnorm(eta)

> i.probit <- irls(died ~ hmo + white,
+                   family = "binomial", link = "probit",
+                   data = medpar)

> summary(i.probit)

Call:
irls(formula = died ~ hmo + white, data = medpar,
     family = "binomial", link = "probit")

Deviance Residuals:
   Min. 1st Qu.  Median    Mean 3rd Qu.    Max.
-0.9268 -0.9268 -0.9223 -0.1002  1.4510  1.5930

Coefficients:
             Estimate     SE     Z        p     LCL     UCL
(Intercept)   -0.5719 0.1184 -4.83 1.37e-06 -0.8040 -0.340
hmo           -0.0073 0.0913 -0.08    0.936 -0.1862  0.172
white          0.1842 0.1233  1.49    0.135 -0.0575  0.426

Null deviance: NA   on  1494 d.f.
Residual deviance: 1920.602  on  1492 d.f.
AIC:   1926.602

Number of optimizer iterations:  3
```

The reader can verify that these results closely approximate the values produced by `glm`, as presented below.

```
> coef(summary(i.glm <- glm(died ~ hmo + white,
+                           family = binomial(probit),
+                           data = medpar)))
```

	Estimate	Std. Error	z value	Pr(>|z|)
(Intercept)	-0.571875132	0.11841182	-4.82954440	1.368458e-06
hmo	-0.007304429	0.09127856	-0.08002349	9.362186e-01
white	0.184209525	0.12330153	1.49397600	1.351819e-01

The same amendment to the grouped logistic function can be made so that one may estimate grouped *probit* and *cloglog* models.

4.8 GLM Negative Binomial Model

The negative binomial is a GLM only if the value of heterogeneity parameter, α, is fixed. In our model, it must be entered into the IRLS estimating algorithm as a constant. It is the only member that is fitted by standard GLM software that has a heterogeneity parameter. It is therefore instructive to see how it can be fitted using the GLM algorithm.

The traditional negative binomial — the model used to accommodate over-dispersed Poisson models — is in fact a non-canonical parameterization of the negative binomial distribution. In order to deal appropriately with over-dispersed Poisson count data, the link used for the negative binomial needs to be the same as that of the Poisson model, namely, the log link. A boundary likelihood ratio test (Hilbe, 2011) may be used to evaluate if α, or $1/\theta$, significantly differs from 0.

The canonical negative binomial link is called the *negative binomial link*: $-\log(1/(\alpha\mu) + 1)$; it can effectively be used to model count data, but not specifically for modelling over-dispersed Poisson counts. Referred to as NB-C, the canonical negative binomial is a member of the GLM family, and can be estimated by several of the foremost GLM applications. It is not, however, part of R's `glm` function, or of the `glm.nb` function from MASS. `glm.nb` provides for a maximum likelihood estimation of θ, the inverse of the heterogeneity parameter α. The *COUNT* package, available on CRAN, has maximum likelihood functions for estimating NB-C (`ml.nbc.r`), as well as NB-2 (`ml.nb2.r`) and NB-1 (`ml.nb1.r`), with the respective heterogeneity parameters provided to have a direct relationship to the excess correlation in the data.

R's `glm` and `glm.nb` functions parameterize the negative binomial heterogeneity parameter, θ, as inversely related to μ, unlike other commercial software applications. `glm` negative binomial coefficients and standard errors may differ a little from other applications, and from how we construct the function, but it is `glm`'s heterogeneity parameter that may cause confusion. For our `irls` function, if α is close to 0, the model is Poisson. For numerical reasons it cannot exactly equal 0. Increasing values of α indicate increasing correlation, or over-dispersion, in the data. To the contrary, `glm`'s and `glm.nb`'s θ indicates a model as Poisson if it approaches infinity. Small values of θ indicate large amounts of over-dispersion in the data. Considerable care must be taken when comparing the results of negative binomial estimation using `glm` and `glm.nb` with that based on other software.

Note also that the `glm` function negative binomial heterogeneity parameter, α, is not estimated; it is entered into the IRLS estimating algorithm as a constant. How does one select the appropriate value for α? The answer for these tools is that the researcher must determine by trial-and-error which value of α results in a model having the Pearson dispersion as close to 1.0 as possible. We recommend using the `ml.nb2` function, located in the *COUNT* package, to determine the maximum likelihood estimate for α, and then use that value for `irls`. One may also use the `glm.nb` function in the *MASS* package, which estimates θ using a full maximum likelihood procedure. Recall that GLMs are one-parameter models; the method does not itself estimate α as a second parameter.

For a comprehensive presentation of the varieties of negative binomial models, and of count models in general; see Hilbe (2011). Table 4.15 provides a stand-alone GLM negative binomial, with a direct relationship of μ and α. Note that the standard errors are produced on the basis of the expected information matrix, which differ slightly from standard errors produced using the observed information matrix. This difference holds only for non-canonically linked models, and is minimal unless there are only a relatively few observations in the model. For this model the parameter estimate standard errors differ beginning with the one-hundred thousandths place.

Now we need three functions that will enable the use of a new link function, that is the link, the inverse link, and the first derivative of the link; and two for the new family, the variance and the joint log-likelihood, again best defined using R's internal functions.

```
> linkFn.log <- function(mu, m, a) log(mu)
> lPrime.log <- function(mu, m, a) 1/mu
> unlink.log <- function(y, eta, m, a) exp(eta)

> variance.negBinomial <- function(mu, m, a) mu + a*mu^2
> jllm.negBinomial <- function(y, mu, m, a) {
+    dnbinom(y, mu = mu, size = 1 / a, log = TRUE)
+ }
```

We will cover explicit estimation of the scale parameter in the next chapter. In the meantime, the maximum likelihood estimate of *alpha* can be determined using the `glm.nb` function, which produces full maximum likelihood estimates of the parameter estimates, including `alpha`.

```
> library(MASS)
> ml.nb <- glm.nb(los ~ hmo + white,
+                     data = medpar)

> summary(ml.nb)

Call:
glm.nb(formula = los ~ hmo + white, data = medpar,
        init.theta = 2.063546582, link = log)

Deviance Residuals:
    Min       1Q    Median       3Q      Max
-2.0870  -0.8490  -0.2657   0.3774   5.5163

Coefficients:
             Estimate Std. Error z value Pr(>|z|)
(Intercept)   2.48102    0.06715  36.947  < 2e-16 ***
hmo          -0.14051    0.05463  -2.572  0.01011 *
white        -0.18971    0.07020  -2.702  0.00689 **
---
Signif. codes:  0 '***' 0.001 '**' 0.01 '*' 0.05 '.' 0.1 ' ' 1

(Dispersion parameter for Negative Binomial(2.0635) family
 taken to be 1)

    Null deviance: 1585.7  on 1494  degrees of freedom
Residual deviance: 1570.7  on 1492  degrees of freedom
AIC: 9706.1

Number of Fisher Scoring iterations: 1


            Theta:  2.0635
        Std. Err.:  0.0893

 2 x log-likelihood:  -9698.0790
```

The standard errors from `glm.nb` are unscaled. `alpha` is the inverse of `theta`, which is displayed and saved in the `glm.nb` function output.

```
> 1 / ml.nb$theta
```

```
[1] 0.4846026
```

We may now use the value of `alpha` in our `irls` function.

```
> irls.nb <- irls(los ~ hmo + white, a = 0.4846026,
```

```
+                     family = "negBinomial", link = "log",
+                     data = medpar)
> nb.summ <- summary(irls.nb)

Call:
irls(formula = los ~ hmo + white, data = medpar,
     family = "negBinomial", link = "log", a = 0.4846026)

Deviance Residuals:
   Min. 1st Qu.  Median    Mean 3rd Qu.    Max.
-2.0870 -0.8490 -0.2657 -0.2385  0.3774  5.5160

Coefficients:
              Estimate     SE     Z         p     LCL      UCL
(Intercept)      2.481 0.0672 36.95 8.23e-299   2.349   2.6126
hmo             -0.141 0.0546 -2.57   0.0101  -0.248  -0.0334
white           -0.190 0.0702 -2.70  0.00689  -0.327  -0.0521

Null deviance: NA   on  1494 d.f.
Residual deviance: 1570.678  on  1492 d.f.
AIC:   9704.079

Number of optimizer iterations:  4
```

These estimates match those produced by R's `glm.nb`. In order to obtain similar results using `glm` function, the user must invert the value of the heterogeneity parameter `alpha` that we assigned to the model `irls` function, or we may obtain it directly from the `glm.nb` function results above.

```
> ml.nb$theta

[1] 2.063547
```

The function command and resultant parameter estimates are given as:

```
> glm.nb <- glm(los ~ hmo + white,
+               data = medpar,
+               family = negative.binomial(2.063547))

> summary(glm.nb)

Call:
glm(formula = los ~ hmo + white,
    family = negative.binomial(2.063547), data = medpar)

Deviance Residuals:
    Min      1Q   Median      3Q      Max
```

```
-2.0870  -0.8490  -0.2657   0.3774   5.5163
```

```
Coefficients:
            Estimate Std. Error t value Pr(>|t|)
(Intercept)  2.48102    0.07738  32.064   <2e-16 ***
hmo         -0.14051    0.06295  -2.232   0.0257 *
white       -0.18971    0.08089  -2.345   0.0191 *
---
Signif. codes:  0 '***' 0.001 '**' 0.01 '*' 0.05 '.' 0.1 ' ' 1
```

```
(Dispersion parameter for Negative Binomial(2.0635) family
 taken to be 1.327754)
```

```
    Null deviance: 1585.7  on 1494  degrees of freedom
Residual deviance: 1570.7  on 1492  degrees of freedom
AIC: 9704.1
```

```
Number of Fisher Scoring iterations: 4
```

It is important to note that for this model, R has estimated the dispersion and scaled the standard error estimates accordingly. Note that

```
> sqrt(summary(glm.nb)$dispersion)
```

```
[1] 1.152282
```

is the (multiplicative) difference between the standard errors of `irls.nb` and `glm.nb`. Here, we scale the values from `irls.nb` using the dispersion estimate supplied by the latter.

```
> nb.summ$coefficients[,"SE"] * sqrt(irls.nb$p.dispersion)
```

```
[1] 0.07737680 0.06294517 0.08089112
```

The match with the standard errors reported in the output directly above seems good.

4.9 Offsets

Note that we added an *offset* to allow fitting rate-parameterized models. A rate parameterization allows for the counts to be apportioned across time intervals or periods as well as for different areas. Most Poisson and negative binomial count models used in research employ offsets to adjust for counts

134 *Methods of Statistical Model Estimation*

being taken over different periods or areas. The basic model assumes that counts are recorded over areas or time periods of unity.

An offset is added to the linear predictor, and is itself not parameterized. That is, no coefficient is estimated; it is assumed that the offset has a slope of 1. In the GLM algorithm, care must be taken not to include the offset into the weighted regression, so at each iteration it is subtracted from the working response. The working response, traditionally given the symbol z, is the response term entered into the weighted lm function, which provides GLM estimates and whose variance-covariance matrix is used for determining the coefficient standard errors. The offset is then added to the linear predictor, which immediately follows the regression. This relationship is clear from viewing the irls code in Section 4.7.

We shall use first use the **heart** data shown below to model the rate parameterized negative binomial. The data consists of senior Canadian patients who have either a Coronary Artery Bypass Graft surgery (CABG) or Percutaneous Transluminal Coronary Angioplasty (PTCA) heart procedure. The response is **death** within 48 hours of hospital admission. Predictors are:

anterior, 1: anterior site damage heart attack; 0: other site damage;
hcabg, 1: previous CABG procedure; 0= previous PTCA procedure;
killip, 1: normal heart; 2: angina; 3: minor heart blockage; 4: heart attack or myocardial infarction;
cases, the number of patients with the same covariate pattern. The offset will be log(cases).

```
> library(msme)
> data(heart)
```

We examine the data to start with.

```
> heart
```

```
   death cases anterior hcabg killip
1     48  1864        0     0      1
2     15   412        0     0      2
3     10    83        0     0      3
4      5    19        0     0      4
5      7    70        0     1      1
6      4    18        0     1      2
7      2     3        0     1      3
8      7    10        0     1      4
9     50  1374        1     0      1
10    39   443        1     0      2
11     9   139        1     0      3
12    10    28        1     0      4
13     5    27        1     1      1
14     3    16        1     1      2
15     2     6        1     1      3
```

The offset must be entered in log form to accord with the log link being used for the negative binomial family. We provide `alpha` with a value of 0.0001 since the true value is close to 0; that is, it is a Poisson model. However, we shall model it as a negative binomial.

```
> heart.nb <- irls(death ~ anterior + hcabg + factor(killip),
+                    a = 0.0001,
+                    offset = log(heart$cases),
+                    family = "negBinomial", link = "log",
+                    data = heart)

> summary(heart.nb)

Call:
irls(formula = death ~ anterior + hcabg + factor(killip),
    data = heart, family = "negBinomial", link = "log",
    offset = log(heart$cases), a = 1e-04)


Deviance Residuals:
    Min.   1st Qu.   Median     Mean   3rd Qu.     Max.
-1.46500 -0.37420  0.08714  0.06542  0.45710  1.59100


Coefficients:
                  Estimate    SE     Z         p       LCL     UCL
(Intercept)        -3.663 0.122 -29.97 2.2e-197 -3.902 -3.423
anterior            0.380 0.140   2.71  0.00672  0.105  0.655
hcabg               1.327 0.205   6.46 1.02e-10  0.925  1.729
factor(killip)2     0.670 0.161   4.17 3.07e-05  0.355  0.985
factor(killip)3     0.997 0.232   4.30 1.67e-05  0.543  1.451
factor(killip)4     2.172 0.241   9.01 1.99e-19  1.700  2.645


Null deviance: NA   on  14 d.f.
Residual deviance: 10.41831   on  9 d.f.
AIC:   82.39765


Number of optimizer iterations:  6
```

We can check our use of the offset against a comparable model fit from R, as follows.

```
> h.glm <- glm(death ~ anterior + hcabg + factor(killip),
+              family = negative.binomial(10000),
+              offset = log(cases),
+              data = heart)
> coef(summary(h.glm))

              Estimate Std. Error    t value    Pr(>|t|)
```

```
(Intercept)      -3.6626879  0.1359945 -26.932624 6.493108e-10
anterior          0.3802929  0.1561427   2.435547 3.763856e-02
hcabg             1.3269729  0.2284759   5.807934 2.569091e-04
factor(killip)2   0.6700463  0.1789052   3.745259 4.588111e-03
factor(killip)3   0.9971516  0.2577766   3.868279 3.798737e-03
factor(killip)4   2.1723303  0.2682073   8.099446 2.005173e-05
```

As before, R has scaled the standard error estimates using its estimate of the dispersion, so we should compare after scaling the standard error estimates from `heart.nb` using

```
> sqrt(heart.nb$p.dispersion)
```

```
[1] 1.112884
```

for example, as follows.

```
> with(heart.nb, se.beta.hat * sqrt(p.dispersion))
```

```
      X(Intercept)         Xanterior             Xhcabg
         0.1359932         0.1561414          0.2284741
Xfactor(killip)2 Xfactor(killip)3 Xfactor(killip)4
         0.1789034         0.2577774          0.2682046
```

The comparison with the output directly above seems reasonable.

4.10 Dispersion, Over- and Under-

A point should be made about grouped binomial, Poisson, and negative binomial models. When used with real data situations, these models are oftentimes over-dispersed. Essentially this means that there is more variation in the data than allowed by the distributional assumptions of the model. In particular, the Poisson model is most usually over-dispersed, which is indicated when the conditional variance exceeds the conditional mean. Theoretically the two statistics should be identical.

For binomial models, over-dispersion is indicated when the dispersion statistics based on deviance or Pearson's Chi2 are greater than unity. For count models, over-dispersion is indicated only when the Pearson-based dispersion is greater than unity. The Pearson dispersion statistic can be obtained following estimation using the `irls` function by:

```
> irls.nb$p.dispersion
```

```
[1] 1.327739
```

This reported model is substantially over-dispersed.

Typically statisticians will scale the standard errors using the Pearson Chi2 dispersion statistic. R provides scaled standard errors in a few different ways. First, scaled standard errors can be obtained for the binomial and Poisson families by using the `quasi-` equivalent family. The `quasibinomial` algorithm sets the standard errors to the values that would be the estimates if the model had a Pearson dispersion of 1, i.e., if there were no excessive correlation in the data. The `quasipoisson` family plays a similar role for the Poisson family. `glm.nb` ignores negative binomial over-dispersion altogether, whereas `glm` with the negative binomial family will estimate the dispersion and scale the standard errors. Alternatively, researchers also employ a robust or sandwich variance estimator to the standard errors, or bootstrap them.

The Pearson and deviance dispersion statistics are obtained by dividing the Pearson Chi2 statistic or residual deviance, respectively, by the model residual degrees of freedom. Both the residual deviance and residual degrees of freedom are displayed in glm summary function output. The Pearson Chi2 statistic must be calculated separately using the code shown later in this section. It is important to keep in mind that over-dispersion in grouped binomial models can be diagnosed using either dispersion statistic, but the deviance dispersion is biased with respect to count models, and should not be used to evaluate possible model over-dispersion (Hilbe, 2011). We discuss this fact in more detail later in the book. That portion of glm output that can be used for assessing the fit of `glm.glog` to the `doll` data appears as

```
> anova(glm.glog)

Analysis of Deviance Table

Model: binomial, link: logit

Response: cbind(deaths, pyears - deaths)

Terms added sequentially (first to last)
```

	Df	Deviance	Resid. Df	Resid. Dev
NULL			9	939.71
smokes	1	29.20	8	910.51
ordered(age)	4	898.41	4	12.10

We can compute the deviance dispersion via

```
> 12.10 / 4

[1] 3.025
```

The Pearson dispersion can be computed from the corresponding Chi2 statistic, as follows.

```
> sum(residuals(glm.glog, type="pearson")^2) /
+     glm.glog$df.residual
```

[1] 2.78052

Note that the deviance and Pearson Chi2 dispersion statistics are quite different. When this occurs with grouped binomial models it indicates that the model may not fit well. In this case, over-dispersion is indicated using either dispersion statistic.

Simulation studies have demonstrated that the Pearson dispersion is the appropriate statistic to use for assessing possibly count model extra-dispersion (Cameron and Trivedi, 1998; Hilbe, 2009, 2011). A true Poisson model is equi-dispersed, with a Pearson dispersion value of approximately 1. Dispersion values over 1 indicate over-dispersion; values under 1 indicate under-dispersion. The negative binomial is typically used to model over-dispersed Poisson data. Under-dispersed Poisson data may be handled by scaling, by use of a hurdle model, by generalized Poisson or by using a generalized negative binomial model. The Pearson dispersion statistic is not displayed in the summary function output following the glm function Poisson model, but it is for the quasipoisson model, which is a Poisson model with standard errors multiplied (scaled) by the Pearson dispersion statistic. To check for Poisson over-dispersion, we advise checking summary function results following the use of the quasipoisson model. Negative binomial over-dispersion may be assessed by calculating the Pearson dispersion statistic for negative binomial models, which differs from the Poisson model by replacing the Poisson with the negative binomial variance function. Recall that the negative binomial variance is defined as $\mu + \alpha\mu^2$, where α is the negative binomial scale parameter when the distribution is parameterized to have a direct relationship with both the fitted value and the amount of correlation or dispersion in the data. A negative binomial model with $\alpha = 0$ is Poisson. Greater values of α indicate greater amounts of over-dispersion in the data.

Readers may use a simple function called P__disp to calculate the Pearson Chi2 and Pearson dispersion statistics following the use of glm or glm.nb functions. It is in both the *COUNT* and *msme* packages on CRAN.

```
> P__disp <- function(x) {
+     pr <- sum(residuals(x, type="pearson")^2)
+     dispersion <- pr / x$df.residual
+     return(c(pearson.chi2 = pr, dispersion = dispersion))
+ }
```

The function can be used directly following estimation of a glm or glm.nb model by typing the name of the model as the sole argument of the function; for example,

```
> P__disp(glm.glog)
```

```
pearson.chi2    dispersion
   11.12208      2.78052
```

4.11   Goodness-of-Fit and Residual Analysis

4.11.1   Goodness-of-Fit

We have touched on assessing the goodness-of-fit in the last section when discussing dispersion. Checking the binomial, Poisson, or negative binomial dispersion statistic, δ, is a first step in assessing whether the model in fact is over-dispersed. There is no clear-cut criterion, however, that specifies a model to be over-dispersed if δ has a given value. The importance of δ differs for a model with few observations compared with a model having a large number of observations, and/or between models having a sizable difference in the number of predictors. For example, a Poisson model having more than 50,000 observations and ten predictors is likely over-dispersed with a δ of 1.1, whereas $\delta = 1.1$ for a model having 30 observations and 2 predictors is not. A boundary likelihood ratio test (see Hilbe, 2011, §7.4.1) can be used to determine if the value of α is significantly greater than 0, i.e., that it is negative binomial rather than Poisson. The p-value of the test is normally based on a Chi2 distribution. However, for this test specifically, the p-value is one-half the probability that a Chi2 value, with one degree of freedom, is greater than the obtained likelihood ratio (LR) statistic. The LR statistic with respect to the boundary likelihood ratio test for Poisson and negative binomial models is given as

$$LR = -2(L_P - L_{NB}) \tag{4.10}$$

The likelihood ratio test is also commonly used to compare nested models whose parameters are estimated using maximum likelihood. The test assumes the independence of observations. It is therefore not appropriate for panel models or any model exhibiting a clustering effect. In any case, the boundary likelihood ratio test evaluates if there is a significant amount of Poisson over-dispersion.

As an example, we now use `irls` to estimate Poisson (`modelP`) and NB2 (`modelNB`) models for the same data. First we need to extend the `irls` function to allow for the Poisson model. We write the following two functions.

```
> jllm.poisson <- function(y, mu, m, a) {
+   dpois(x = y, lambda = mu, log = TRUE)
+ }
> variance.poisson <- function(mu, m, a) mu
```

We can now easily fit the Poisson model to the **heart** data.

```
> heart.p <- irls(death ~ anterior + hcabg + factor(killip),
+                   offset = log(heart$cases),
+                   family = "poisson", link = "log",
+                   data = heart)
```

The boundary likelihood ratio statistic and p-value may be calculated as:

```
> LR <- -2*(heart.p$loglik - heart.nb$loglik)
> pchisq(LR, 1, lower.tail = FALSE)/2
```

```
[1] 0.5
```

Over-dispersion tends to result in predictor p-values based on z or Wald statistics appearing as significant when in fact they are not. In the presence of over-dispersion, estimates of standard errors are deflated, resulting in inflated z statistics, which produce lower p-values. p-values lower than 0.05 are traditionally considered to be significant. Therefore, care must be taken to accommodate any over-dispersion in the data.

Under-dispersion may also occur in binomial and count models. Scaling may adjust for under-dispersion as well. Binary response models cannot themselves be extra-dispersed; i.e., either under- or over-dispersed. However, a binary model may be implicitly over-dispersed if, when converted to a binomial format, it exhibits over-dispersion (see, Hilbe, 2009).

Care must be taken to determine if evident over-dispersion, e.g., as judged by the value of δ, is in fact only apparent and not real over-dispersion. If $\delta > 1$ for a given model, then it may be that adding a specific predictor or predictors to the model results in a re-modelling of the data to have $\delta \sim 1$. In this case the over-dispersion was not real, but only apparent. Likewise, if deleting extreme outliers in the data, or creating one or more interaction terms, or amending the scale of one or more of the predictors, or changing the link function for the model results in δ being close to 1, the original model was only apparently over-dispersed. It is vital to test for apparent over-dispersion prior to declaring a model over-dispersed and taking corrective action.

Traditionally the deviance statistic was regarded as a comparative indicator of better fit. Of two nested models, the one with the lowest deviance statistic was regarded as the better fitted model. The AIC (Akaike Information Criterion) and BIC (Bayesian Information Criterion) statistics are now the most favored comparative fit statistics. AIC and BIC statistics can be used to compare the fit to data across non-nested as well as nested models.

Models having a lower AIC or BIC statistic are preferred to those having higher values. A table evaluating the degree of difference in AIC statistics and the corresponding degree of confidence one has in selecting that model over another is given by Hilbe (2011). The form of AIC used most frequently in R software is

$$AIC = -2(L - k) \tag{4.11}$$

with L indicating the value of the model log-likelihood and k specifying the number of predictors in the model, including the intercept.

It should be noted that a variety of alternative AIC-type statistics have been devised, all employing adjustment by both the number of predictors and of observations in the model.

The most commonly used form of BIC statistic appears to be Schwarz Criterion or the Schwarz BIC, which is the original formulation of the statistic (Schwarz, 1978). It appears as

$$BIC = -2L + k \times \log(n) \qquad (4.12)$$

4.11.2   Residual Analysis

The foremost value in the analysis of residuals rests in the ability of the viewer to see patterns, emphases, outliers, and so forth that may not easily be discerned from statistics alone. We may know the mean and variance of a given distribution of counts, but not realize from those figures any abnormalities that may be present in the data. The counts may be clustered in one or more areas of the distribution – a fact that we could not know, or at least easily know, other than by viewing it.

Statisticians have developed a number of residuals, nearly all of which are extensions of the basic *raw* or *response residual*, which is defined as a value of the response or dependent variable less the expected or predicted mean, μ. This relationship, for a single observation, has been given various expressions; e.g., $y - E(y)$, $y - \hat{y}$, or $y - \hat{\mu}$. We have employed the latter form, and will continue it here.

Residual analysis is used by analysts to (1) detect outliers, or groups of outliers, in the data, and to (2) check the distributional assumptions of the model. If the shape of the residuals varies considerably from what would be expected given a well-fitted model, we can in general conclude that the model poorly fits the data. In this case, we may need to amend the link function, the variance, or perhaps even the basic type of model itself.

Within the context of GLM modelling, there are five levels or types of residual, each type being in general a more detailed adjustment to the basic response residual. A second level of residuals is called *modified residuals*. These include Pearson and deviance residuals, which are modified to provide a reasonable estimate of the conditional variance of the response variable, y.

The *Pearson residual* is defined as

$$R_P = \frac{y - \hat{\mu}}{\sqrt{V(\hat{\mu})}} \qquad (4.13)$$

which is nothing more than the raw residual standardized by a gross estimate of the standard deviation of y. Note that the sum of squared Pearson residuals is the model Pearson Chi2 statistic. Therefore, the Pearson residual can be thought of as a per-observation contribution to the Pearson Chi2 statistic. Unfortunately the residual can be markedly skewed in non-linear models.

The *deviance residual* is

$$R_D = sgn\,(y - \widehat{\mu})\,\sqrt{deviance} \tag{4.14}$$

Like the Pearson residual, the model deviance statistic is defined as the sum of squared deviance residuals. Deviance residuals measure each observation's contribution to the model deviance. Note should be given that we constructed the deviance statistic in the `irls` function by summing the squared deviance statistics. By so doing the algorithm creates both the deviance and deviance residuals in a single line. We did the same for the Pearson Chi2 statistic.

Standardizing the Pearson or deviance residuals normalizes them to a standard deviation of 1.0, and creates a third level of residual. The deviance residual appears to normalize the residual better than the Pearson residual, and statisticians have as a consequence developed a number of tests based on it. It is now the preferred statistic for residual analysis.

Standardization is achieved by dividing the residual by $\sqrt{1-h}$, where h is the *hat* matrix diagonal statistic. The standardized deviance appears as

$$R_{SD} = \frac{R_D}{\sqrt{(1-h)}} \tag{4.15}$$

h is a measure of the influence of an observation to the model, and is often also referred to as the leverage.

$$h = V(\widehat{\mu}) \times S_p^2 \tag{4.16}$$

S_p is the standard error of the prediction, or var($\widehat{\mu}$). Rs `glm` and `irls` functions both abstract the *hat* statistic from the model matrix, X, with `irls` abstracting it as a QR decomposition of X. var($\widehat{\mu}$) may be obtained this way by abstracting $X(X'WX)^{-1} X'$ from h. Conversely, h may be created by adding the two $W^{1/2}$ matrices to the extremes of var($\widehat{\mu}$).

Standardized Pearson and deviance residuals are sometimes scaled to further assist in better effecting normality by including a dispersion statistic within the denominator radical of the standardized residual. This residual statistic is called a *studentized residual*, and is one of several complex residuals constructed using additional statistics. The studentized deviance residual appears as:

$$R_{Stud} = \frac{R_D}{\sqrt{\phi\,(1-h)}} \tag{4.17}$$

4.12 Weights

For full maximum likelihood and IRLS functions, e.g., `glm`, the user may employ frequency or prior weights into the model by use of the weights argument. Frequency weights are used when there are multiple instances of the same pattern of predictor values. Grouped logistic regression may also be estimated using a frequency weight approach. Prior weights may be used with any of the glm and standard MLE models, although they are rarely used with Poisson and negative binomial models.

Sampling weights are employed using the same format as for prior weights. However, sampling weights are expressed as fractions or decimals. The software recognizes the non-integer nature of the values given in using the weights argument, and calculates sampling weights in place of frequency or prior weights. A sampling weight is normally used in survey studies to adjust for sampling from unequal sub-populations. The weight is lower for smaller sub-populations, and higher for larger ones. The weight itself is calculated as the inverse of the probability of being in the selected sample. It therefore will be a decimal, unless the probability is zero or one, which is unlikely. Caveats have been raised about the use of sampling weights except for prediction, see Gelman and Hill (2007) and Aitkin et al. (2010) for a discussion.

4.13 Conclusion

GLM models rest at the foundation of a variety of more enhanced modelling techniques, including random, fixed, and mixed effects models, hierarchical models, generalized estimating equations (GEE), Bayesian modelling, generalized additive models, and so forth. GEE employs the GLM algorithm for estimating parameters, but adjusts the standard errors of the model at each iteration using a correlation matrix supplied by either the software or the user. GEE models are also estimated using IRLS methodology. Random effects and most other panel models need to use more complex estimation methods for their solutions, which we discuss further in Chapter 6.

GLM and GEE models provide for a wide range of modelling capabilities. GLMs can be used to estimate most single parameter models. The modular approach we discussed in this chapter is an efficient way to handle algorithms with multiple groups of related functions. However, it should be kept in mind that the `irls` function we developed has limitations, particularly when the data is ill-formed or considerably unbalanced. R is sometimes fickle in how it handles these types of data situations. Our purpose in designing `irls` is not to supplant R's `glm` function, but rather to show the logic of constructing IRLS-based functions.

4.14 Exercises

1. The model that we fitted to the `doll` data had substantial over-dispersion. What might cause that? What potential remedies are there? Try them.

2. Use the `heart` data to show that the results from `irls` agree with the earlier `irls_logit` (see Section 4.6).

3. Write a stand-alone `irls`-type Poisson model with canonical log link. Offer the capability of having offsets in your model. Verify your results, both with and without offsets, using both R's `glm` function and the `irls` function from the text.

4. Extend the `irls` function to encompass the Poisson family. Verify your results, both with and without offset, using R's `glm` function.

5. Display the generic equation for the exponential family of distributions, giving θ as the location parameter, ϕ as the scale, and $c()$ as the normalization function. Indicate how each term in the PDF relates to the IRLS algorithm that is used to estimate models that are members of the family of generalized linear models.

6. Extra-dispersion:

   (a) How can extra-dispersion be identified in a grouped binomial model, as well as in Poisson and negative binomial models?

   (b) What is the difference between the negative binomial *theta* parameter used with R's `glm` and `glm.nb` functions and the negative binomial *alpha* parameter used for the negative binomial functions in this text? What are the advantages in using the *alpha* parameterization?

   (c) Can negative binomial models adjust for Poisson under-dispersion? Why or why not? Can they be adjusted for negative binomial under-dispersion? Why or why not?

7. APR's habitual approach to modelling the `ufc` dataset has been to assume that the tree heights are missing completely at random. Fit a model to assess the evidence against this assumption.

5

Maximum Likelihood Estimation

5.1 Introduction

In Chapter 3 we presented an overview of how to construct a maximum likelihood function for linear regression. A fairly complete model was presented, with code to demonstrate how to maximize the log-likelihood, determine residuals, display a summary function, and allow for post-estimation statistics. The IRLS algorithm, which provides maximum likelihood estimates for a limited set of models, was discussed in Chapter 4. We constructed several different types of functions for estimating standard GLM models, and then developed a modular umbrella `irls` function for estimation of binomial, Poisson, negative binomial, gamma, and inverse Gaussian models. All models discussed in Chapter 4 have only a single parameter to be estimated.

We now shall expand our previous discussion of maximum likelihood estimation (MLE) to demonstrate how the criterion may be used to estimate a much larger range of both GLM and non-GLM models. Examples of non-GLM models that use MLE for the solution of parameter estimates and associated standard errors include non-linear models, two-parameter models, categorical response models, models with a mixture of distributions, GLM-based models with an likelihood function amended to handle distributional violations, e.g., zero-inflated Poisson, models in which the data are correlated, and survival models. The two foremost keys determining if a model can be straightforwardly estimated using MLE are (1) the independence of model observations, and (2) if there is a closed-form solution to the estimating equation of the model. The first of these two criteria is often violated, as we shall discuss in Chapter 6.

MLE, as the name indicates, maximizes the log-likelihood, equivalently it involves determining the values of the parameters for which the derivative of the model likelihood (equivalently, the model log-likelihood) is zero.

$$\left.\frac{\partial L}{\partial \beta}\right|_{\beta=\hat{\beta}} = 0 \qquad (5.1)$$

Estimates of the model parameters are the solutions of this estimating equation. The second derivative of the log-likelihood with respect to β is the Hessian matrix from which model standard errors are calculated.

5.2 MLE for GLM

We now develop the code that can be used to fit generalized linear models using maximum likelihood. The order of development of the functions is reasonably arbitrary. We start with logistic regression, and the code starts with the sample. Our goal here is to develop an algorithm that is more general than the IRLS algorithm developed in the previous chapter. We start, however, with a simple example.

5.2.1 The Log-Likelihood

Our goal is to find the values of parameters that maximize the joint log-likelihood of the sample of data. There are many different equally good ways that the code to perform this task could be written. Our goal is clarity of exposition, and then flexibility, rather than code efficiency or stability, and we code accordingly. This means that we write more functions than seem to be needed for the problem at hand. However, this detail will reap benefits later on. As in previous chapters, we will use S3 classes.

The joint log-likelihoods of a set of observations y conditional on its fitted values $\hat{y}$ for logistic regression can be written as

```
> jll.bernoulli <- function(y, y.hat, ...) {
+    dbinom(x = y, size = 1, prob = y.hat, log = TRUE)
+ }
```

we know that this probably will not be the only log-likelihood that we define, so we will also define a generic function called jll.

```
> jll <- function(y, y.hat, ...) UseMethod("jll")
```

Now, as in previous chapters, when R evaluates this function it will start by identifying the class of the first argument. R will then use the method that corresponds to the first match that it can find in the vector of class names. We will need to ensure that it finds "bernoulli".

Our jll function will report a vector of logged probabilities. In order to use this quite general function we will need another function that sums the result, given the following inputs: the predictor variables coded as a model matrix X, the response variable as a vector y, and some candidate parameter estimates b.hat.

```
> Sjll <- function(b.hat, X, y, offset = 0, ...) {
+    y.hat <- predict(y, b.hat, X, offset)
+    sum(jll(y, y.hat, ...))
+ }
```

N.B.: The astute reader will wonder why we have separated the functions for the calculation of the joint log-likelihood and its summation. We elected to separate these functions because doing so allows us to re-use the jll function when the time comes to compute the deviance residuals.

The Sjll function still needs to compute the fitted values as a function of the observations and the predictor variables.

```
> predict.expFamily <- function(object, b.hat, X, offset = 0) {
+    lin.pred <- as.matrix(X) %*% b.hat + offset
+    y.hat <- unlink(object, lin.pred)
+    return(y.hat)
+ }
```

Note that we carry the y argument in order to tell R which unlink method to use. An alternative would be to copy the class information to one of the other arguments. We can use the same idea for the unlink function as was defined in Chapter 4 for the irls function. We do not need to code the link function or its derivative. We name the new unlink function logit1 in order to distinguish it from the logit function written earlier; this version can only be used for Bernoulli regression.

```
> unlink <- function(y, eta, ...) UseMethod("unlink")
> unlink.logit1 <- function(y, eta, ...) 1 / (1 + exp(-eta))
```

Now we have all the paraphernalia in place that we will use to compute the log-likelihood of the parameter estimates conditional on the model and the data. It is useful to test the functions with simple cases. For example, we can use the following invented data and parameter estimates.

```
> y <- c(1,0,0,1,1,1,0,1)
> X <- as.matrix(cbind(1, 1:8))
> beta.hat <- c(0,1)
```

In order to take full advantage of the S3 classes that we constructed earlier in the chapter, we need to specify the family, the link function, and be sure that the correct predict function is used. We pass all this information by means of the class of the response variable.

```
> class(y) <- c("bernoulli","logit1","expFamily")
```

We can now evaluate the summed joint log-likelihood at the parameter estimates.

```
> Sjll(beta.hat, X, y)

[1] -12.51736
```

The output provides us with two bits of good news: first, our functions seem to work sensibly, and second, the log-likelihood can be evaluated at the parameter estimates that we chose. This means that we can use them as a start point for our future optimization efforts. Our next challenge is to maximize that function, or more precisely, to find the values of the parameter estimates for which the output of the function is maximized.

5.2.2 Parameter Estimation

We next write a wrapper that will find the values of parameters that maximize a function, conditional on predictor variables. We will use the following function to maximize the joint log-likelihood. It is very similar to the approach that we took in Chapter 3. We are writing a separate function rather than just performing the optimization within the main function because as we shall soon see, we will want to perform the optimization twice; once for the specified model and once for the null model.

```
> maximize <- function(start, f, X, y, offset = 0, ...) {
+    optim(par = start,
+          fn = f,
+          X = X,
+          y = y,
+          offset = offset,
+          method = "BFGS",
+          control = list(
+             fnscale = -1,
+             reltol = 1e-16,
+             maxit = 10000),
+          hessian = TRUE,
+          ...
+       )
+ }
```

We now have functions that can be used to obtain logistic regression MLEs from data. We might as well use the previous guess as a start point; we now know that the likelihood can be evaluated here. We obtain the parameter estimates as

```
> test <- maximize(beta.hat, Sjll, X, y)
> test$par
```

```
[1] -0.1721367  0.1552640
```

We can compare these estimates with output from R's `glm` function, but we will defer that comparison until a little later in the chapter. For the moment it is comforting that we obtain parameter estimates at all. Of course, we could use `irls` for this particular problem, but we are building a model-fitting tool that will fit more general models than IRLS can easily handle[1].

Our next challenge is to obtain other quantities from the fitted model that can be used to assess the model, both in terms of how well our assumptions are met and in comparison with other models. We start by examining the residuals.

5.2.3 Residuals

Numerous different types of residuals have been defined for use in GLMs, for example, Hardin and Hilbe (2007) list nine different kinds, many of which could also be modified, standardized, studentized, or adjusted. Our goal is not to be encyclopedic, so we will focus on just the deviance residuals.

The deviance residual for an observation evaluated at the fitted value of $\hat{\mu}$ is defined generally as the contribution to the deviance of the observation.

$$\hat{d}_i = \sqrt{2l(y, y) - 2l(\hat{\mu}, y)} \tag{5.2}$$

The deviance residual then takes on the sign of the difference between y and $\hat{\mu}$, so that it is negative if the fitted value is larger than the observed.

For example, in the binomial case, the formula for the i-th deviance residual given observation y_i, predicted value $\hat{\mu}_i$, and count m_i, can be derived as:

$$\hat{d}_i = \sqrt{2y_i \ln\left(\frac{y_i}{\hat{\mu}_i}\right) + 2(m_i - y_i) \ln\left(\frac{m_i - y_i}{m_i - \hat{\mu}_i}\right)} \tag{5.3}$$

noting that

$$y_i \times \ln(y_i)\big|_{y_i=0} = 0 \tag{5.4}$$

It is common for books about GLM to provide these derived equations for each member of the exponential family. The use of these equations allows efficient computation of the residuals, but at the same time it detracts from the underlying unity of the technique by splintering the exponential family into a suite of seeming special cases. Hence we provide only one function to compute the deviance residuals, and it uses the joint log-likelihood.

In words, the magnitude of the deviance residual for an observation is the square root of twice the difference between the joint log-likelihood evaluated at the datum and evaluated at the fitted value. Here, "evaluated at the datum" means that the predicted value is y *and* the observation is y. In code, if $y = 1$, and $\hat{y} = 0.5$ then we would write something like:

```
> sqrt(2 * dbinom(x = 1, size = 1, prob = 1, log = TRUE) -
+         dbinom(x = 1, size = 1, prob = 0.5, log = TRUE))
```

---

[1]The reader may wish to compare the execution times of these two algorithms, using `system.time`.

[1] 0.8325546

For our code, we now take advantage of our earlier structure. We need to calculate the predicted values using the parameter estimates, the design matrix, and the unlink function, which we can do using the same `predict.expFamily` function as before.

```
> devianceResiduals <- function(y, b.hat, X, offset = 0, ...) {
+     y.hat <- predict(y, b.hat, X, offset)
+     sign(y - y.hat) *
+         sqrt(2 * (jll(y, y, ...)) - jll(y, y.hat, ...))
+ }
```

The utility of this function is that it can be used to produce the deviance residuals and, as we shall see, the deviance, for models that represent any member of the exponential family, or indeed for any model family for which deviance residuals are considered informative. We merely need to write an appropriate `jll` function.

5.2.4   Deviance

Now we can report the deviance of the model simply as the sum of the squares of the residuals defined in the previous section. For example, the deviance of the model that we fitted earlier is calculated by

```
> sum(devianceResiduals(y, test$par, X)^2)
```

[1] 5.177151

We can also find the null deviance in a quite general way: compute the sum of the squares of the deviance residuals, fitted at the null model. In this case, it would be

```
> fit.null <- maximize(0.5, Sjll, 1, y)
> sum(devianceResiduals(y, fit.null$par, 1)^2)
```

[1] 5.292506

We can check whether we are getting these deviance values right by comparing them with output from R's own `glm` function. Our version of X already has the intercept, so we should omit it here.

```
> glm(y ~ X[,-1], family=binomial)
```

```
Call:  glm(formula = y ~ X[, -1], family = binomial)
```

```
Coefficients:
```

```
(Intercept)      X[, -1]
  -0.1721        0.1553
```

```
Degrees of Freedom: 7 Total (i.e. Null);  6 Residual
Null Deviance:              10.59
Residual Deviance: 10.35           AIC: 14.35
```

This output shows us that our residual and null deviance values agree with those provided by R, and also provides a check on the parameter estimates computed in the previous section. We are confident that, so far, our code is reasonable. To further simplify the process of using the function, we will add an option to automate the selection of starting points for the optimizer.

5.2.5   Initial Values

As in previous chapters, we need to find initial values for the optimizer. It is easy enough if we know that we will always be working with the raw data, but becomes complicated when we use different link functions. In order to provide a general solution, we will write an S3 generic function and a default version of it that simply grabs the coefficients from an ordinary least-squares fit of the predictors to the response, minus the offset.

```
> kickStart <- function(y, X, offset)
+                               UseMethod("kickStart")
> kickStart.default <- function(y, X, offset = 0) {
+    coef(lm(I(y - offset) ~ X - 1))
+ }
```

Our experimentation suggests that this function will suffice for a binary response variable. Later versions will, for example, grab the coefficients from an ordinary least-squares fit of the predictor variables to the log of the response minus the offset.

5.2.6   Printing the Object

We will want to be able to report the output of the model in some convenient way. We could write our own print function, but since we are writing a GLM model, it makes sense to use as much of the existing R infrastructure as we can. One strategy is to declare the class of the returned object as being both *msme* and *glm*, the former so that we can write special functions for it as needed, and the latter so that we can take advantage of any existing *glm* functions for convenience. This strategy uses the inheritance quality of S3 classes.

However, this alone will not be sufficient, because the `print.glm` function will assume that certain pieces of information will be available. In order to be able to use this function, we must either (i) examine a glm object, identify all of its components, and make sure that our object has the same components, or, preferably, (ii) examine the `print.glm` function, and be sure that all of the components that it expects to find are provided. The latter is easier for now.

```
> print.glm

function (x, digits = max(3, getOption("digits") - 3), ...)
{
    cat("\nCall:  ", paste(deparse(x$call), sep = "\n",
        collapse = "\n"), "\n\n", sep = "")
    if (length(coef(x))) {
        cat("Coefficients")
        if (is.character(co <- x$contrasts))
            cat("  [contrasts: ", apply(cbind(names(co), co),
                1L, paste, collapse = "="), "]")
        cat(":\n")
        print.default(format(x$coefficients, digits = digits),
            print.gap = 2, quote = FALSE)
    }
    else cat("No coefficients\n\n")
    cat("\nDegrees of Freedom:", x$df.null,
        "Total (i.e. Null); ", x$df.residual, "Residual\n")
    if (nzchar(mess <- naprint(x$na.action)))
        cat("  (", mess, ")\n", sep = "")
    cat("Null Deviance:\t    ",
        format(signif(x$null.deviance, digits)),
        "\nResidual Deviance:", format(signif(x$deviance,
        digits)), "\tAIC:", format(signif(x$aic, digits)), "\n")
    invisible(x)
}
<bytecode: 0x104ed9ea0>
<environment: namespace:stats>
```

Given an object x of class *glm*, this function assumes that it will be able to locate the named components `call`, `coefficients`, `df.null`, `df.residual`, `null.deviance`, `deviance`, and `aic`. If we also provide `contrasts` and `na.action` then it will also report those. The `print.glm` function also assumes that `coef(x)` will return the coefficients. We should check what that means by examining the function. Some experimentation leads to the following:

```
> getAnywhere(coef.default)

A single object matching 'coef.default' was found
It was found in the following places
  registered S3 method for coef from namespace stats
```

```
      namespace:stats
with value
```

```
function (object, ...)
object$coefficients
<bytecode: 0x234f694>
<environment: namespace:stats>
```

which suggests that we just need to be sure that our object includes `coefficients`. We will use the outcome of this exercise to guide the design of the object that our new `glm` function will return.

5.2.7 GLM Function

We now link these various steps together in a single function, much as we did with linear regression in Chapter 3 and iteratively re-weighted least squares in Chapter 4. The function follows the main themes of: handle the input, prepare the model infrastructure, check for missing data, choose starting values, estimate the parameters, check for validity, compute relevant statistics, and report.

```
> ml_glm <- function(formula,
+                    data,
+                    family,
+                    link,
+                    offset = 0,
+                    start = NULL,
+                    verbose = FALSE,
+                    ...) {
+
+ ### Handle the input
+    mf <- model.frame(formula, data)
+    y <- model.response(mf, "numeric")
+
+ ### Prepare model infrastructure
+    class(y) <- c(family, link, "expFamily")
+    X <- model.matrix(formula, data = data)
+
+ ### Check for missing data.  Stop if any.
+    if (any(is.na(cbind(y, X)))) stop("Some data missing!")
+
+ ### Initialize the search, if needed
+    if (is.null(start))  start <- kickStart(y, X, offset)
+
+ ### Maximize the joint log-likelihood
+    fit <- maximize(start, Sjll, X, y, offset, ...)
```

```
+
+ ### Check for optim failure and report and stop
+   if (verbose | fit$convergence > 0)  print(fit)
+
+ ### Extract and compute quantities of interest
+   beta.hat <- fit$par
+   se.beta.hat <- sqrt(diag(solve(-fit$hessian)))
+   residuals <- devianceResiduals(y, beta.hat, X, offset, ...)
+
+ ### Fit null model and determine null deviance
+   fit.null <- maximize(mean(y), Sjll, 1, y, offset, ...)
+   null.deviance <-
+     sum(devianceResiduals(y, fit.null$par, 1, offset, ...)^2)
+
+ ### Report the results, with the needs of print.glm in mind
+   results <- list(fit = fit,
+                   X = X,
+                   y = y,
+                   call = match.call(),
+                   obs = length(y),
+                   df.null = length(y) - 1,
+                   df.residual = length(y) - length(beta.hat),
+                   deviance = sum(residuals^2),
+                   null.deviance = null.deviance,
+                   residuals = residuals,
+                   coefficients = beta.hat,
+                   se.beta.hat = se.beta.hat,
+                   aic = - 2 * fit$val + 2 * length(beta.hat),
+                   i = fit$counts[1])
+
+ ### Use (new) msme class and glm class
+   class(results) <- c("msme","glm")
+   return(results)
+ }
```

An example of the use of the function follows. First, attach the *msme* package and obtain the `medpar` data.

```
> library(msme)
> data(medpar)
```

Then fit a simple model using the formula interface.

```
> mort.glm <- ml_glm(died ~ hmo + white,
+                     data = medpar,
+                     family = "bernoulli",
+                     link = "logit1")
```

Now we can print the model in the same neat arrangement that has already been designed for the S3 class of *glm* objects, as follows.

```
> mort.glm
```

```
Call:  ml_glm(formula = died ~ hmo + white, data = medpar,
    family = "bernoulli", link = "logit1")
```

```
Coefficients:
X(Intercept)          Xhmo        Xwhite
    -0.92619       -0.01225       0.30339
```

```
Degrees of Freedom: 1494 Total (i.e. Null);   1492 Residual
Null Deviance:              1923
Residual Deviance: 1921           AIC: 1927
```

Had we not arranged to use the existing `print.glm` function, then the result would have been to print the list of objects, piece by piece. That output would still be informative, but not nearly as useful.

Components of the object are also available to examine. For example, if we would like to know more about the status of the conclusion of the optimization routine, then we type the following.

```
> mort.glm$fit
```

```
$par
X(Intercept)          Xhmo        Xwhite
 -0.92618625   -0.01224651    0.30338726
```

```
$value
[1] -960.301
```

```
$counts
function gradient
      51       14
```

```
$convergence
[1] 0
```

```
$message
NULL
```

```
$hessian
              X(Intercept)       Xhmo      Xwhite
X(Intercept)    -336.47134  -53.81772  -310.67613
Xhmo             -53.81772  -53.81772   -51.39212
Xwhite          -310.67613  -51.39212  -310.67613
```

The last piece of infrastructure that we need is a function that provides useful statistics computed from the fitted model. For example, we would like to print a coefficient table. We wrote a custom `summary.irls` method for this purpose in the previous chapter. Happily, the function also works with objects developed in this chapter, as called below, so long as it (`summary.msme`, Section 4.7.1) and the following residuals function are both loaded. Loading the book's package is the easiest approach.

```
> residuals.msme <- function(object,
+                                 type = c("deviance","standard"),
+                                 ...) {
+   type <- match.arg(type)
+   if (type == "standard") {
+     object$residuals / sqrt(1 - diag(hatvalues(object)))
+   } else {
+     object$residuals
+   }
+ }
```

Moreover, we also wanted to add confidence intervals to the default summary output, unlike `print.summary.glm`, which displays the coefficients, standard errors, z, and *p*-values, but not the confidence intervals, which must be called for separately using the `confint` or, for Wald confidence intervals, `confint.default`.

```
> summary(mort.glm)

Call:
ml_glm(formula = died ~ hmo + white, data = medpar,
    family = "bernoulli", link = "logit1")

Deviance Residuals:
   Min. 1st Qu.  Median    Mean 3rd Qu.    Max.
-0.9268 -0.9268 -0.9222 -0.1002  1.4510  1.5930


Coefficients:
             Estimate      SE        Z         p       LCL      UCL
(Intercept) -0.92619  0.1974 -4.69215 2.703e-06 -1.31307 -0.5393
hmo         -0.01225  0.1489 -0.08223    0.9345 -0.30414  0.2796
white        0.30339  0.2052  1.47864    0.1392 -0.09876  0.7055

Null deviance: 1922.865  on  1494 d.f.
Residual deviance: 1920.602  on  1492 d.f.
AIC:  1926.602

Number of iterations:  51
```

All the fundamental pieces are now in place to fit logistic regression models. We could write a large collection of functions to add further functionality, for example, functions to extract information such as the Pearson Chi2 for assessing dispersion, functions to perform model comparison, and the like, but our goal is not to replace the existing, excellent R functionality, but rather to use it as a framework for demonstrating statistical models.

5.2.8 Fitting for a New Family

Having invested all the effort into getting a reasonable infrastructure, it is now straightforward to extend our model to fit other kinds of GLM. For example, the Poisson distribution is a member of the exponential family. In order to use our function to perform Poisson regression, we only need a few pieces: the Poisson joint log-likelihood, an unlink function that corresponds to the log link, and a new function to prescribe the starting point for the parameters when the log link function will be used. Examples of such functions follow.

First, we write a joint log-likelihood for the Poisson distribution, for example,

```
> jll.poisson <- function(y, y.hat, ...) {
+   dpois(y, lambda = y.hat, log = TRUE)
+ }
```

The unlink function for the log link is simply **exp**. Note that again we pass y as the first argument so that the appropriate method will be selected from the class of y.

```
> unlink.log <- function(y, eta, m=1, a=1) exp(eta)
```

Finally we need a method to select the start point when the predictors will be filtered through a log link function. A simple solution is to get parameter estimates from a linear regression of the log of y against the predictors.

```
> kickStart.log <- function(y, X, offset = 0) {
+   coef(lm(I((log(y + 0.1) - offset) ~ X - 1)))
+ }
```

Having declared these three functions, we can now use our **ml_glm** function to fit a Poisson regression model, as follows.

```
> ml.poi <- ml_glm(los ~ hmo + white,
+                   family = "poisson",
+                   link = "log",
+                   data = medpar)
```

The summary output, below, compares well with the output from R's **glm** function.

```
> summary(ml.poi)

Call:
ml_glm(formula = los ~ hmo + white, data = medpar,
    family = "poisson", link = "log")

Deviance Residuals:
   Min. 1st Qu.  Median    Mean 3rd Qu.    Max.
-4.1200 -1.7490 -0.6213 -0.3073  0.9434 18.9500


Coefficients:
             Estimate     SE      Z        p      LCL      UCL
(Intercept)    2.4823 0.02589 95.890      0   2.4315  2.53299
hmo           -0.1416 0.02373 -5.966 2.431e-09 -0.1881 -0.09507
white         -0.1909 0.02728 -6.998   2.6e-12 -0.2444 -0.13742

Null deviance: 8901.134  on  1494 d.f.
Residual deviance: 8812.942  on  1492 d.f.
AIC:  14534.09


Number of iterations:  58

> summary(glm(los ~ hmo + white,
+             family = poisson,
+             data = medpar))

Call:
glm(formula = los ~ hmo + white, family = poisson,
    data = medpar)

Deviance Residuals:
    Min      1Q   Median      3Q      Max
-4.1197  -1.7487  -0.6213  0.9434  18.9477

Coefficients:
             Estimate Std. Error z value Pr(>|z|)
(Intercept)  2.48225    0.02589  95.890  < 2e-16 ***
hmo         -0.14158    0.02373  -5.966 2.43e-09 ***
white       -0.19089    0.02728  -6.998 2.60e-12 ***
---
Signif. codes:  0 '***' 0.001 '**' 0.01 '*' 0.05 '.' 0.1 ' ' 1

(Dispersion parameter for poisson family taken to be 1)

    Null deviance: 8901.1  on 1494  degrees of freedom
```

```
Residual deviance: 8812.9  on 1492  degrees of freedom
AIC: 14534
```

```
Number of Fisher Scoring iterations: 5
```

This exercise demonstrates a number of important elements: the elegance and unity of GLM, the utility of S3 classes, and the importance of planning ahead in software design. For example, we defined the residual deviance in terms of the deviance residuals, and defined the deviance residuals using the joint log-likelihood, which also appeared in the objective function. Therefore our switch of PDF from Bernoulli to Poisson changed all the relevant portions of the model. Had we used efficient, stable code to compute the deviance or the residuals then we would have to completely rewrite it for the new model.

For further illustration, we now use maximum likelihood to fit a zero-truncated Poisson model (ZTP). We start by writing a function for the joint log-likelihood. The PDF for the ZTP is constructed as the PDF for the Poisson distribution scaled by the probability that a random Poisson number is not zero.

$$f(x; \theta | x > 0) = \frac{\theta^x e^{-\theta}}{x!} \frac{1}{1 - e^{-\theta}} \tag{5.5}$$

We can write this in R as follows:

```
> jll.ztp <- function(y, y.hat, ...)
+   dpois(y, lambda = y.hat, log = TRUE) - log(1 - exp(-y.hat))
```

We can now fit the ZTP model using the same function as we did before.

```
> ml.ztp <- ml_glm(los ~ hmo + white,
+                  family = "ztp",
+                  link = "log",
+                  data = medpar)
```

The summary output, below, compares favorably with results from other software (not shown here).

```
> summary(ml.ztp)
```

```
Call:
ml_glm(formula = los ~ hmo + white, data = medpar,
    family = "ztp", link = "log")
```

```
Deviance Residuals:
    Min. 1st Qu.  Median    Mean 3rd Qu.    Max.    NA's
 -2.8990 -1.9170 -1.0170 -0.3807  0.9496 13.2700     944
```

Coefficients:

	Estimate	SE	Z	p	LCL	UCL
(Intercept)	2.482	0.0259	95.89	0	2.432	2.5330
hmo	-0.142	0.0237	-5.97	2.41e-09	-0.188	-0.0952
white	-0.191	0.0273	-7.00	2.57e-12	-0.244	-0.1375

```
Null deviance: NaN  on  1494 d.f.
Residual deviance: NaN  on  1492 d.f.
AIC:   14533.89

Number of optimizer iterations:  53
```

5.3 Two-Parameter MLE

We now extend the set of models into two-parameter members of the exponential family. We will start with the negative binomial distribution with log link for the mean and the scale parameters. That is, we will construct the following model and its multivariate extensions.

$$y_i \sim NB(\exp(\beta_0 + \beta_1 x_{1i}), \exp(\delta_0 + \delta_1 x_{2i})) \tag{5.6}$$

Much of the work that was done in the previous section can be re-used, but not all of it. Some functions will need to be extended, and some will need to be replaced. Here we will see further examples of the limitations of S3 classes[2], which will force us to develop rather clumsy functions to handle the breadth of models that we wish to use. As before, we start by defining the log-likelihood that we will try to maximize.

5.3.1 The Log-Likelihood

The two-parameter likelihoods that we shall deal with in this section differ from the previous likelihoods only in that they have an extra parameter, the scale of the density. This means that we will need to develop new functions that will accommodate the estimation of the parameters that are associated with the scale. We start with a method for the negative binomial joint log-likelihood, as follows.

```
> jll2.negBinomial <- function(y, y.hat, scale, ...) {
+    dnbinom(y,
+            mu = y.hat,
+            size = 1 / scale,
```

---

[2]Or, more accurately, the limitations of the authors' ability to use S3 classes.

```
+            log = TRUE)
+ }
```

Note that by setting `size = 1 / scale`, we have reparameterized the negative binomial as suggested by Cameron and Trivedi (1998, p. 71–73), Winkelmann (2008, p. 134–135), and Hilbe (2011, p. 9–10) so that the variance is directly proportional to the mean, instead of inversely proportional.

We will also need a new generic function, similar to that we used earlier. We provide it again here for clarity.

```
> jll2 <- function(y, y.hat, scale, ...) UseMethod("jll2")
```

Our joint log-likelihood must be summed, and the parameters of the model must be passed to it in some organized way. The summation is straightforward, as before, but the handling of the predictor variables requires some planning. Ultimately we would like to be able to fit models for which both parameters are functions of some predictor variables. Hence, we will want to specify two design matrices: X1 and X2. For convenience, we may prefer to use one matrix that has a specific structure: the first p columns correspond to the mean parameter and the balance of the columns to the scale parameter. We will use this approach.

Therefore the sum of the log-likelihoods can be arranged as follows.

```
> Sjll2 <- function(b.hat, X, y, p, offset = 0, ...) {
+   y.hat <- predict(y, b.hat[1:p], X[,1:p], offset)
+   scale.hat <- predict_s(y, b.hat[-(1:p)], X[,-(1:p)])
+   sum(jll2(y, y.hat, scale.hat, ...))
+ }
```

In this function we split the design matrix into two portions: the first p columns, and the rest. Using the `predict.expFamily` function, we multiply the former with the first p parameter estimates, and apply the unlink function that we elected for the mean — here it will be the `exp` function (see Section 5.2.8). Then using the `predict_s` function, we multiply the remaining columns by the remaining parameters and apply the unlink function that was nominated for the scale. Here it will also be the `exp` function (see below).

The `predict_s` function will convert the scale's parameter estimates and design matrix into a linear predictor, and then apply an unlink function, if any is nominated.

```
> predict_s <- function(y, b.hat, X) {
+   lin.pred <- as.matrix(X) %*% b.hat
+   scale <- unlink_s(y, lin.pred)
+   return(scale)
+ }
```

It would be elegant to be able to use a new `predict` method, but all the class information is carried by y, and the `predict` method is dedicated to the mean function. It cannot be used for both.

Finally, we need to write an unlink function that we can use to handle the fitting and reporting of the model for the scale parameter. Here we will assume that a log link is used, for the moment. As before, we write the method for the class of interest and also a generic method.

```
> unlink_s.log_s <- function(y, eta) exp(eta)
> unlink_s <- function(y, eta) UseMethod("unlink_s")
```

We now have sufficient functions to compute the log-likelihood, evaluated at given parameter estimates, and conditional on the model and the data. For example, we can use the following invented data and parameter estimates.

```
> y <- c(1,0,0,1,1,3,0,9)
> X <- as.matrix(cbind(1, 1:8, 1, 1:8))
> beta.hat <- c(0,1,0,1)
```

As before, we need to pass the class information that the S3 system will use via y.

```
> class(y) <- c("negBinomial","log","log_s","expFamily")
```

Now we can try to evaluate the summed joint log-likelihood at the nominated parameter values.

```
> Sjll2(beta.hat, X, y, 2)
```

```
[1] -29.30543
```

We note that the function seems to work, and that it can be evaluated at the parameter values. We will next think about finding the values of the parameter estimates that maximize the joint log-likelihood.

5.3.2   Parameter Estimation

The `maximize` function that we will re-use needs to pass the argument p, which is the number of columns of X that are relevant to the mean parameter. We take advantage of the `dots` argument which passes arguments to internal functions. During earlier drafts we tried to pass p as an attribute of the matrix X but were sadly unsuccessful, as the matrix shed the attribute during processing.

As before, it is prudent to test our code along the way.

```
> test <- maximize(beta.hat, Sjll2, X, y, p = 2)
> test$par
```

```
[1]  -2.3543758   0.5223218  -3.5761943 -20.0290300
```

Whether these estimates are reasonable remains to be seen.

5.3.3 Deviance and Deviance Residuals

The deviance residuals are constructed in the same way as before: using the definition of the deviance, and determining the contribution of each observation to that deviance, multiplied by an appropriate sign.

However, the calculation raises a different challenge. Recall that the definition of the deviance is twice the difference between the joint log-likelihood evaluated at the data and evaluated at the fitted value. The open question is: what should happen to the scale parameter during this operation? It cannot be set to the observation. This is the ancillary/nuisance parameter conundrum. The consensus seems to be that the scale should be kept fixed at its MLE for the purposes of determining the deviance of the model. The use of two prediction functions, one for the fitted values and one for the scale, should be familiar by now.

A further complication is that the deviance, and therefore the deviance residuals, have to be scaled by a function of the dispersion. Whilst we were working with the binomial and poisson models, we could ignore this issue because the dispersion for those two models is assumed to be 1. This is also true for the negative binomial distribution; however, for distributions that we will be fitting later in the chapter, the dispersion will vary.

```
> getDispersion <- function(y, scale) UseMethod("getDispersion")
> getDispersion.negBinomial <- function(y, scale) 1

> devianceResiduals2 <- function(y, b.hat, X, p, offset = 0) {
+    y.hat <- predict(y, b.hat[1:p], X[,1:p], offset)
+    scale <- predict_s(y, b.hat[-(1:p)], X[,-(1:p)])
+    sign(y - y.hat) *
+      sqrt(2 * getDispersion(y, scale) *
+        (    jll2(y, y,      scale) -
+             jll2(y, y.hat, scale)   ))
+ }
```

We can now evaluate the deviance residuals for any two-parameter distribution for which we can provide a suitable joint log-likelihood and an appropriate dispersion — just so long as we are convinced that the deviance residuals have some useful interpretation.

Finally, we can again evaluate the deviance as the sum of the squares of the deviance residuals. For our current model, that would be

```
> sum(devianceResiduals2(y, test$par, X, 2)^2)
```

```
[1] 12.30863
```

We now make a quick comparison with the glm.nb function from the *MASS* package (Venables and Ripley, 2010). Here is the output for the negative binomial model that has only a single parameter for the scale.

```
> library(MASS)

> glm.nb(y ~ X[,2])

Call:  glm.nb(formula = y ~ X[, 2], init.theta = 3.008721266,
              link = log)

Coefficients:
(Intercept)        X[, 2]
   -1.9810        0.4582

Degrees of Freedom: 7 Total (i.e. Null);   6 Residual
Null Deviance:                   16.68
Residual Deviance: 8.415            AIC: 30.51
```

From our code, we get

```
> test <- maximize(beta.hat[1:3], Sjl12, X[,1:3], y, p = 2)
> test$par

[1] -1.9810214  0.4582397 -1.1015178
```

In comparing the scale parameter estimates, recall that those from our code should be transformed in the following way: they should be exponentiated (as we were using a log-link) and they should be inverted (as we had re-parameterized the negative binomial distribution). We correct as follows, and the result matches the value **init.theta** reported by **glm.nb**.

```
> exp(-test$par[3])

[1] 3.008729
```

The comparison of the parameter estimates seems very satisfactory, as does the comparison of the deviance.

```
> sum(devianceResiduals2(y, test$par, X[,1:3], 2)^2)

[1] 8.415389
```

Computing the null deviance proceeds along the same lines as before, but here again we have to make some decision about what to do with the scale parameter in computing the null deviance. As before, the standard seems to be to use the MLE of the scale conditional on the fitted model. Hence, the null deviance of the model would be calculated from the fitted values evaluated at the null fit, but the scale values evaluated at the model fit.

We can now move with some confidence to develop more of the modelling infrastructure. The worst challenges are behind us, but some challenges remain. We now need to find initial values for the scale parameter estimates as well.

5.3.4   Initial Values

The automatic determination of initial values for the optimizer for all possible models that could be fit takes us beyond the scope of our code. Hence it is probable that the reader will be able to discover datasets and models for which our code does not work. Our goal is to demonstrate, rather than to be comprehensive. We can take advantage of our earlier code that determines start points for the parameters associated with the fitted value, and use a simplification for the start points of the scale parameters.

```
if (is.null(start)) {
  start <- c(kickStart(y, X1, offset),
             1,
             rep(0, ncol(X) - p - 1))
  names(start) <- c(colnames(X1), colnames(X2))
}
```

Note that we name the start values using the column names of the two design matrices. These names will be carried through the analysis and facilitate easy reporting.

For the current example we will need to use the `kickStart` method that finds initial parameter estimate for the linear predictor when the log link will be used.

5.3.5   Printing and Summarizing the Object

We retain this brief sub-section in order to make explicit the point that we can entirely re-use the code written for printing or summarizing the fitted object in the one-parameter case, just so long as we endure that the fitted object contains the appropriate components. That is, we do not have to rewrite any of these functions because the earlier versions are perfectly suitable.

5.3.6   GLM Function

We now link these various steps together in a single function. The function follows the main themes of: handle the input, prepare the model infrastructure, check for missing data, choose starting values, estimate the parameters, check for validity, compute relevant statistics, and report.

```
> ml_glm2 <- function(formula1,
+                     formula2 = ~1, data,
+                     family,
+                     mean.link,
+                     scale.link,
+                     offset = 0,
+                     start = NULL,
```

```
+                              verbose = FALSE) {
+
+ ### Handle the input
+    mf <- model.frame(formula1, data)
+    y <- model.response(mf, "numeric")
+
+ ### Prepare model infrastructure
+    class(y) <- c(family, mean.link, scale.link, "expFamily")
+    X1 <- model.matrix(formula1, data = data)
+    X2 <- model.matrix(formula2, data = data)
+    colnames(X2) <- paste(colnames(X2), "_s", sep="")
+    p <- ncol(X1)
+    X <- cbind(X1, X2)
+
+ ### Check for missing data.  Stop if any.
+    if (any(is.na(cbind(y, X)))) stop("Some data are missing!")
+
+ ### Initialize the search
+    if (is.null(start)) {
+      start <- c(kickStart(y, X1, offset),
+                 1, # Shameless hack
+                 rep(0, ncol(X) - p - 1))
+      names(start) <- c(colnames(X1), colnames(X2))
+    }
+
+ ### Maximize the joint log likelihood
+    fit <- maximize(start, Sjll2, X, y, offset, p = p)
+
+ ### Check for optim failure and report and stop
+    if (verbose | fit$convergence > 0)  print(fit)
+
+ ### Extract and compute quantities of interest
+    beta.hat <- fit$par
+    se.beta.hat <- sqrt(diag(solve(-fit$hessian)))
+    residuals <- devianceResiduals2(y, beta.hat, X, p, offset)
+
+ #### Deviance residuals for null
+    fit.null <- maximize(c(mean(y), 1),
+                         Sjll2,
+                         X[,c(1,p+1)],  y, offset, p = 1)
+    null.deviance <-
+      sum(devianceResiduals2(y,
+                             c(fit.null$par[1], fit$par[p+1]),
+                             X[, c(1,p+1)],
+                             1,
```

```
+                                     offset)^2)
+
+ ### Report the results, with the needs of print.glm in mind
+    results <- list(fit = fit,
+                    loglike = fit$val,
+                    X = X,
+                    y = y,
+                    p = p,
+                    call = match.call(),
+                    obs = length(y),
+                    df.null = length(y) - 2,
+                    df.residual = length(y) - length(beta.hat),
+                    deviance = sum(residuals^2),
+                    null.deviance = null.deviance,
+                    residuals = residuals,
+                    coefficients = beta.hat,
+                    se.beta.hat = se.beta.hat,
+                    aic = - 2 * fit$val + 2 * length(beta.hat),
+                    offset = offset,
+                    i = fit$counts[1])
+    class(results) <- c("msme","glm")
+    return(results)
+ }
```

The model is run as follows.

```
> test.3.g <- ml_glm2(los ~ hmo + white,
+                     formula2 = ~1,
+                     data = medpar,
+                     family = "negBinomial",
+                     mean.link = "log",
+                     scale.link = "log_s")
```

A print of the model can be obtained by

```
> test.3.g

Call:  ml_glm2(formula1 = los ~ hmo + white, formula2 = ~1,
    data = medpar, family = "negBinomial",
    mean.link = "log", scale.link = "log_s")

Coefficients:
  (Intercept)            hmo          white   (Intercept)_s
       2.4810        -0.1405        -0.1897         -0.7244


Degrees of Freedom: 1493 Total (i.e. Null);  1491 Residual
```

```
Null Deviance:              1586
Residual Deviance: 1571            AIC: 9706
```

and a summary by

```
> summary(test.3.g, dig = 2)
```

```
Call:
ml_glm2(formula1 = los ~ hmo + white, formula2 = ~1,
    data = medpar, family = "negBinomial",
    mean.link = "log", scale.link = "log_s")
```

```
Deviance Residuals:
   Min. 1st Qu.  Median    Mean 3rd Qu.     Max.
-2.0870 -0.8490 -0.2657 -0.2385  0.3774   5.5160
```

```
Coefficients:
               Estimate    SE     Z         p   LCL     UCL
(Intercept)       2.48 0.067  37.0 3.7e-299  2.35   2.613
hmo              -0.14 0.055  -2.6     0.01 -0.25  -0.033
white            -0.19 0.070  -2.7   0.0069 -0.33  -0.052
(Intercept)_s    -0.72 0.043 -16.7  6.7e-63 -0.81  -0.640
```

```
Null deviance: 1585.658  on  1493 d.f.
Residual deviance: 1570.678  on  1491 d.f.
AIC:  9706.079
```

```
Number of iterations:   80
```

Note that the formatting of the above table of coefficients was displayed for the purposes of fitting easily on the page. The actual results have coefficients and confidence intervals presented to four decimal places, standard error to five, z at three and p at six. More decimal places may be obtained using the post-estimation coefficients option. The negative binomial scale parameter estimate is displayed in log form, -0.7244. Exponentiating the estimate results in the scale parameter having the inverse of its standard form, namely $1/2.0635 = 0.4846$. Recall that we chose a different parameterization than that used by glm.nb. Exponentiating the coefficients results in their parameterization as incidence rate ratios.

```
> test.3.g$coefficients
```

```
  (Intercept)           hmo          white (Intercept)_s
    2.4810189    -0.1405090     -0.1897072    -0.7244261
```

```
> exp(-0.7244261)
```

[1] 0.4846026

We may generate predicted or fitted model values, and summarize, them with a single line of code:

```
> str(with(test.3.g,
+          predict(y, coefficients[1:p], X[,1:p])))

 num [1:1495, 1] 9.89 8.59 8.59 9.89 9.89 ...
 - attr(*, "dimnames")=List of 2
 ..$ : chr [1:1495] "1" "2" "3" "4" ...
 ..$ : NULL
```

We have also provided a number of post-estimation statistics in the results section of the code, including deviance residuals, coefficients and standard errors, residual degree of freedom, and so forth. For example, the model log-likelihood may be displayed as

```
> test.3.g$loglike
```

[1] -4849.039

We suggest observing how the saved values were created and saved. You may add others that were required for the types of study with which you are involved. Using saved values from the function you develop is easier than having to re-create them each time.

We can compare the output with a comparable model fitted using the glm.nb function from *MASS*. The degrees of freedom differ because our model counts the estimated scale as a parameter.

```
> summary(glm.nb(los ~ hmo + white, data = medpar))

Call:
glm.nb(formula = los ~ hmo + white, data = medpar,
    init.theta = 2.063546582, link = log)

Deviance Residuals:
    Min       1Q   Median       3Q      Max
-2.0870  -0.8490  -0.2657   0.3774   5.5163

Coefficients:
            Estimate Std. Error z value Pr(>|z|)
(Intercept)  2.48102    0.06715  36.947  < 2e-16 ***
hmo         -0.14051    0.05463  -2.572  0.01011 *
white       -0.18971    0.07020  -2.702  0.00689 **
---
Signif. codes:  0 '***' 0.001 '**' 0.01 '*' 0.05 '.' 0.1 ' ' 1
```

```
(Dispersion parameter for Negative Binomial(2.0635) family
 taken to be 1)
```

```
    Null deviance: 1585.7  on 1494  degrees of freedom
Residual deviance: 1570.7  on 1492  degrees of freedom
AIC: 9706.1
```

```
Number of Fisher Scoring iterations: 1
```

```
              Theta:   2.0635
          Std. Err.:   0.0893
```

```
 2 x log-likelihood:   -9698.0790
```

The comparison seems quite satisfactory. We may quickly access the standard-form value of the model scale parameter. With the value of the scale parameter given in **glm.nb** as theta, or

```
> coef(test.3.g)[4]
```

```
(Intercept)_s
  -0.7244261
```

we may invert and exponentiate it to standard form with a single line,

```
> exp(-coef(test.3.g)[4])
```

```
(Intercept)_s
   2.063547
```

The standard errors of the model coefficients may be reported as

```
> test.3.g$se.beta.hat[1:3]
```

```
(Intercept)           hmo         white
 0.06711191   0.05465329   0.07022468
```

but the standard errors of the IRR must be calculated using the *delta method*,

$$SE_{IRR} = s_{\hat{\beta}} \times \exp \hat{\beta} \tag{5.7}$$

Refer to Hilbe (2011, p. 23) for details. They may be calculated using the statistics saved in the *ml_glm2* object.

```
> with(test.3.g, se.beta.hat * exp(coefficients))[2:3]
```

```
      hmo       white
0.04748911 0.05808995
```

We reiterate that we do not consider our function suitable for production work. Even though it may fit a broader range of models than `glm.nb`, because of the scale parameterization, it is not as robust. When the data are extreme, or when using offsets, and an offset has the same value as the response, the function may have difficulty converging. Moreover, we have not allowed for the users to employ prior weights when modelling, although giving this capability to the function is not difficult. Use the function with these caveats in mind, or better, enhance the function.

5.3.7 Building on the Model

The model can be extended to mimic the output of other software, for example Stata, by the addition and use of more unlink functions for the scale. Note that in this case, for convenience, we have conflated two issues that could also be considered distinctly: the scale upon which the scale parameter is fit (e.g., raw, log, inverse), and the scale upon which it is reported. These scales need not be the same, but we felt that distinguishing them would add needlessly to the complication, and therefore the code, of the book without contributing much to the information. Three most commonly used unlink functions are:

```
> unlink_s.log_s <- function(y, eta) exp(eta)
> unlink_s.identity_s <- function(y, eta) eta
> unlink_s.inverse_s <- function(y, eta) 1 / eta
```

`ml_glm2` allows the user to parameterize the scale parameter. When applied to the negative binomial scale, or heterogeneity parameter, the model is known as a heterogeneous negative binomial, or NB-H (Hilbe, 2011; Greene, 2012). R does not otherwise have this model, or function, in its scope to our knowledge, although it may be accommodated in one of the many community-written packages. For an example, we may parameterize both model predictors `hmo` and `white` by assigning them to `formula2`. Parameterizing the scale allows the user to help determine which predictors most influence any over-dispersion (or under-dispersion) in the data.

```
> test.3a.g <- ml_glm2(los ~ hmo + white,
+                       formula2 = ~ white + hmo,
+                       data = medpar,
+                       family = "negBinomial",
+                       mean.link = "log",
+                       scale.link = "log_s")
> summary(test.3a.g, dig = 2)

Call:
ml_glm2(formula1 = los ~ hmo + white, formula2 = ~white + hmo,
```

```
        data = medpar, family = "negBinomial", mean.link = "log",
        scale.link = "log_s")
```

Deviance Residuals:
```
   Min. 1st Qu.  Median    Mean 3rd Qu.    Max.
-2.0800 -0.9203 -0.2630 -0.2383  0.3729  5.4510
```

Coefficients (all in linear predictor):
```
               Estimate     SE      Z         p    LCL      UCL
(Intercept)      2.4788 0.0691 35.885 5.18e-282  2.343  2.61414
hmo             -0.1403 0.0511 -2.746   0.00602 -0.240 -0.04018
white           -0.1873 0.0720 -2.599   0.00934 -0.328 -0.04607
(Intercept)_s   -0.6345 0.1397 -4.542  5.56e-06 -0.908 -0.36072
white_s         -0.0634 0.1471 -0.431     0.667 -0.352  0.22490
hmo_s           -0.2432 0.1279 -1.902    0.0572 -0.494  0.00746
```

```
Null deviance: 1477.833  on  1493 d.f.
Residual deviance: 1572.623  on  1489 d.f.
AIC:  9706.119
```

Number of optimizer iterations: 83

```
> test.3a.g <- ml_glm2(los ~ hmo + white,
+                      formula2 = ~ white + hmo,
+                      data = medpar,
+                      family = "negBinomial",
+                      mean.link = "log",
+                      scale.link = "log_s")
> summary(test.3a.g, dig = 2)
```

Call:
```
ml_glm2(formula1 = los ~ hmo + white, formula2 = ~white + hmo,
    data = medpar, family = "negBinomial", mean.link = "log",
    scale.link = "log_s")
```

Deviance Residuals:
```
   Min. 1st Qu.  Median    Mean 3rd Qu.    Max.
-2.0800 -0.9203 -0.2630 -0.2383  0.3729  5.4510
```

Coefficients:
```
               Estimate    SE     Z        p   LCL     UCL
(Intercept)       2.479 0.069 35.89 5.2e-282  2.34  2.6141
hmo              -0.140 0.051 -2.75    0.006 -0.24 -0.0402
white            -0.187 0.072 -2.60   0.0093 -0.33 -0.0461
(Intercept)_s    -0.635 0.140 -4.54  5.6e-06 -0.91 -0.3607
```

```
white_s        -0.063 0.147 -0.43     0.67 -0.35  0.2249
hmo_s          -0.243 0.128 -1.90     0.057 -0.49  0.0075
```

Null deviance: 1477.833 on 1493 d.f.
Residual deviance: 1572.623 on 1489 d.f.
AIC: 9706.119

Number of iterations: 67

Exponentiation of the coefficients of the fully scaled NB-H model above may be calculated as

```
> exp(coef(test.3a.g)[2:3])
```

```
     hmo    white
0.869086 0.829208
```

As previously mentioned, these are known as *incidence rate ratios* (IRR).

We can compare the fits of the two models using their AIC values. Here we do not see strong evidence that the parameterized scale makes much of a difference to the fit, as the AIC values are very close, and that of the more complex model is higher.

```
> test.3.g$aic
```

```
[1] 9706.079
```

```
> test.3a.g$aic
```

```
[1] 9706.119
```

5.3.8 Fitting for a New Family

We are now in a position in which we can extend the coverage of our model with ease. For example, we may wish to use the function to fit a normal regression model that mimics heteroskedasticity. We would do so by writing the joint log-likelihood function and the dispersion function for the normal distribution, as follows.

```
> jll2.normal <- function(y, y.hat, scale, ...) {
+    dnorm(y,
+          mean = y.hat,
+          sd = scale, log = TRUE)
+ }
> getDispersion.normal <- function(y, scale) scale^2
> unlink.identity <- function(y, eta) eta
```

Now we can use the same function to fit a familiar model.

```
> data(ufc)

> ufc <- na.omit(ufc)
> test.1.g <- ml_glm2(height.m ~ dbh.cm,
+                     formula2 = ~1,
+                     data = ufc,
+                     family = "normal",
+                     mean.link = "identity",
+                     scale.link = "log_s")

> summary(test.1.g)

Call:
ml_glm2(formula1 = height.m ~ dbh.cm, formula2 = ~1, data = ufc,
    family = "normal", mean.link = "identity",
    scale.link = "log_s")

Deviance Residuals:
   Min. 1st Qu.  Median   Mean 3rd Qu.    Max.
-33.530  -2.862   0.132  0.000   2.851  13.320



Coefficients:
               Estimate      SE     Z          p     LCL     UCL
(Intercept)     12.6757 0.56262 22.53 2.121e-112 11.5730 13.7784
dbh.cm           0.3126 0.01384 22.58 6.372e-113  0.2855  0.3397
(Intercept)_s    1.5950 0.03576 44.60          0  1.5250  1.6651

Null deviance: 21885.77  on   389 d.f.
Residual deviance: 9497.697  on   388 d.f.
AIC:   2362.938

Number of iterations:   47
```

The output from R's lm function follows.

```
> test.1.lm <- lm(height.m ~ dbh.cm, data = ufc)

> summary(test.1.lm)

Call:
lm(formula = height.m ~ dbh.cm, data = ufc)

Residuals:
    Min      1Q  Median      3Q     Max
-33.526  -2.862   0.132   2.851  13.321
```

```
Coefficients:
            Estimate Std. Error t value Pr(>|t|)
(Intercept) 12.67570    0.56406   22.47   <2e-16 ***
dbh.cm       0.31259    0.01388   22.52   <2e-16 ***
---
Signif. codes:  0 '***' 0.001 '**' 0.01 '*' 0.05 '.' 0.1 ' ' 1
```

Residual standard error: 4.941 on 389 degrees of freedom
Multiple R-squared: 0.566,   Adjusted R-squared: 0.5649
F-statistic: 507.4 on 1 and 389 DF,   p-value: < 2.2e-16

A quick comparison of the scale parameter provides:

```
> exp(test.1.g$coefficients[3])

(Intercept)_s
    4.928569

> summary(test.1.lm)$sigma

[1] 4.941222
```

which is reasonable (recall that the ML estimate is downwardly biased).

We can now extend the model as follows. For example, we can make the variance a function of the diameter.

```
> test.2.g <- ml_glm2(height.m ~ dbh.cm,
+                      formula2 = ~ dbh.cm,
+                      data = ufc,
+                      family = "normal",
+                      mean.link = "identity",
+                      scale.link = "log_s")
```

Then, using the `alrt` function developed earlier (Section 3.4.9), we can compare the fits of the two candidate models, given the following function to extract the likelihoods.

```
> logLik.msme <- function(object, ...) {
+    val <- object$fit$value
+    attr(val, "nall") <- nrow(object$X)
+    attr(val, "nobs") <- nrow(object$X)
+    attr(val, "df") <- length(object$fit$par)
+    class(val) <- "logLik"
+    val
+ }

> alrt(test.1.g, test.2.g)
```

```
LL of model 1:  -1178.469  df:  3
LL of model 2:  -989.2109  df:  4
Difference:  189.258  df:  1
p-value against H_0: no difference between models  0
```

This output suggests that the variation of the heights could change as a function of the tree diameter, which is a biologically reasonable conclusion. These models can be fit more robustly using the gls function of the *nlme* package.

The nbinomial function in the *msme* package provides an alternative and enhancement to glm.nb, with the same model summary output as glm.nb, but also the addition of Pearson residuals, the Pearson Chi2 statistic, and the dispersion statistic. The dispersion parameter is included in the predictor list with standard errors and confidence intervals. We provide the traditional direct parameterization of the dispersion parameter in the default output, and the optional inverse parameterization used in glm.nb. nbinomial also allows parameterization of the dispersion parameter. This procedure is called heterogeneous negative binomial regression in the literature (see, e.g., Hilbe, 2011). Read the nbinomial help file for further information.

5.4   Exercises

1. List the major distributional criteria for maximum likelihood estimation. What are the consequences when each criterion is violated?

2. Rewrite the ml_glm function so that it returns the log-likelihood as loglike. Also write an extractor function.

3. Write a function to extract the Pearson residuals and Pearson Chi2 statistic.

4. Write the needed functions for using ml_glm for exponential regression. This includes functions for the mean, variance, link, inverse link, derivative of the link, deviance, and log-likelihood.

5. Write the needed functions for using ml_glm for grouped binomial regression. A caveat: be certain to take the binomial denominator, m, into consideration, which is the number of trials for a given pattern of covariates. This will need to be passed to the appropriate functions.

6. Write the needed functions for using ml_glm2 for gamma regression.

7. Write the needed methods to allow you to call AIC on a fitted *msme* model. Do this in two ways, with one using AIC.default.

8. Extend the two-parameter ML function to accommodate three-parameter families.

6

Panel Data

6.1 What Is a Panel Model?

The term *panel model* refers to a general set of models aimed at understanding longitudinal, clustered, and nested data. Essentially, though, the term relates to how data are collected. Panel models are collected on observations that are clustered or belong to groups, or on observations over a given period of time. The majority of models that are aimed to estimate the parameters of longitudinal or clustered data are panel models.

Unlike the models we have thus far discussed, which assume that observations in the model are independent of one another, the observations in panel models are not assumed to be independent. Specifically, observations that share a common panel are assumed to be correlated with one another. However, the panels of observations are considered to be independent, conditional on the model. If the panels are considered as independent groups of observations, then maximum likelihood techniques can oftentimes be used for their estimation, particularly if the model is not complex. Other methods are also used to estimate more complicated panel models; e.g., expectation–maximization (EM) and quadrature. In any case, an adjustment must be taken to account for the observations within panels. Typically it is assumed that the within-panel observations are normally distributed with a mean of zero and unknown variance of σ^2.

Table 6.1 displays three panels from a longitudinal German health economics study. This dataset has been used for examples in several leading texts on count models. The original data has unequal sized panels, ranging from one to five observations per panel. We have refashioned the data so that there are 1600 panels comprising 5 observations each for a total of 8000 observations. The five observations per panel each represent a single year of information regarding the number of doctor visits made by patients in the study (docvis). Other variables in the data include id, the panel identifier, year (1984–1988), gender (female = 1; male = 0), age (25–64), and outwork (patient out of work during most of the year = 1; working = 0). The panel sample sizes are equal throughout the data; such data are called *balanced*.

Recall we stated that the data in Table 6.1 was selected so that panel sizes would be equal. The original data comprises panels of unequal sizes, with some patients observed for only a single year, others for two years, and up to five

TABLE 6.1
German health economics study (3 panels only).

id	year	docvis	female	age	outwork
86	1984	2	1	32	0
86	1985	6	1	33	0
86	1986	6	1	34	0
86	1987	6	1	35	0
86	1988	1	1	36	0
104	1984	0	0	46	0
104	1985	0	0	47	0
104	1986	0	0	48	0
104	1987	0	0	49	0
104	1988	2	0	50	0
1330	1984	13	0	28	0
1330	1985	0	0	29	1
1330	1986	2	0	30	1
1330	1987	0	0	31	0
1330	1988	2	0	32	1

years. We selected only those panels that had all five years of data, discarding the remainder. A number of statistical procedures require equal balanced data for estimation purposes.

Table 6.2 displays 32 observations from the 1993 U.S. national Medicare in-patient hospital data, referred to as the Medpar data. The data are appropriately named medpar. We have used this data before in earlier chapters, ignoring the fact that the observations are taken from patients at a number of different hospitals or providers. Note that the provider numbers are associated with unequal numbers of observations. Of the 32 observations displayed in Table 6.2, 5 are associated with provider 030067, 1 with 030068, 19 with 030069, 4 with 030073, and 3 with 030078. The 03 in the provider number designates the state of Arizona; the other numbers specify a hospital within that state. The data represent the number of patients experiencing a specific disease, or more technically a diagnostic related group (DRG) of medical conditions. Hospital 030067 had only 5 patients hospitalized for this condition in 1993. None were members of a Health Maintenance Organization (HMO), all classified themselves as Caucasian (white), and none were hospitalized 10 days or more. LOS is an acronym for *Length of Stay*.

If we were asked to model the count of days a patient is hospitalized los, conditional on hmo and white, we would likely employ a negative binomial model. The response variable, los, is a count. Given the extra correlation in the data due to the panel structure of the data, the data will be over-dispersed. A Poisson model will therefore be inappropriate. In fact, it is possible that the negative binomial model will also be over-dispersed due to the same reason.

TABLE 6.2

From the *medpar* data.

provnum	los	hmo	white
030067	4	0	1
030067	1	0	1
030067	2	0	1
030067	6	0	1
030067	9	0	1
030068	2	0	1
030069	12	0	1
030069	3	0	1
030069	1	0	1
030069	19	0	1
030069	10	0	1
030069	6	0	1
030069	15	0	1
030069	6	0	1
030069	13	0	1
030069	12	0	1
030069	8	0	1
030069	22	0	1
030069	3	0	1
030069	5	0	1
030069	1	0	1
030069	3	0	1
030069	2	0	1
030069	4	0	1
030069	15	0	1
030073	18	0	0
030073	44	0	0
030073	9	0	0
030073	16	0	0
030078	18	0	0
030078	21	0	0
030078	16	0	0

As discussed in the previous two chapters, we may *prima facie* assess count
model over-dispersion by checking the value of the Pearson dispersion statistic,
which can be determined by using the `P__disp` function that we designed for
use following `glm` or `glm.nb` (see Section 4.10).

```
> library(msme)
```

```
> data(medpar)
```

```
> library(MASS)
> nbtest1 <- glm.nb(los ~ hmo + white, data = medpar)
> P__disp(nbtest1)                # dispersion

pearson.chi2    dispersion
  1980.986080     1.327739

> 1/nbtest1$theta               # alpha

[1] 0.4846026
```

the value of 0.4846 above is the negative binomial scale parameterized to
be directly related to μ and to the correlation in the data. It is the standard
manner of expressing the negative binomial scale, or heterogeneity parameter.
Again, it is the inverse of *theta*, which is displayed in glm.nb and glm output as
the negative binomial dispersion parameter. Here *theta* has a value of 2.0635.
The inverse is 0.4846.

Using ml_glm2 from the *msme* package (Section 5.3.6), the model results
in

```
> nbtest2 <- ml_glm2(los ~ hmo + white,
+                     formula2 = ~1,
+                     data = medpar,
+                     family = "negBinomial",
+                     mean.link = "log",
+                     scale.link = "inverse_s")

> summary(nbtest2)

Call:
ml_glm2(formula1 = los ~ hmo + white, formula2 = ~1,
    data = medpar, family = "negBinomial",
    mean.link = "log", scale.link = "inverse_s")

Deviance Residuals:
   Min.  1st Qu.  Median    Mean  3rd Qu.    Max.
-2.0870  -0.8490 -0.2657 -0.2385   0.3774  5.5160

Coefficients:
                Estimate     SE      Z          p     LCL      UCL
(Intercept)        2.481 0.0671  36.97  3.69e-299   2.349   2.6126
hmo               -0.141 0.0547  -2.57     0.0101  -0.248  -0.0334
white             -0.190 0.0702  -2.70     0.0069  -0.327  -0.0521
(Intercept)_s      2.064 0.0893  23.11  3.89e-118   1.889   2.2386

Null deviance: 1585.659  on  1493 d.f.
Residual deviance: 1570.678  on  1491 d.f.
```

```
AIC:  9706.079

Number of optimizer iterations:  66
```

Notice that the `ml_glm2` function parameterizes the negative bi-nomial scale as directly related to model correlation. By using the `scale.link = "inverse_s"` argument, the algorithm inverts `theta` and dis-plays `alpha` together with its appropriate standard errors, z-statistic, p-value, and confidence intervals.

Moderate negative binomial over-dispersion is indicated by the estimated dispersion parameter from `P__disp`, 1.3278. The extra correlation that results in the data from the panel structure may be reduced by incorporating a pa-rameter that accounts for the within-hospital correlation of observations. It may be that staying in different hospitals makes no difference for the length of stay. Two common approaches for handing such data are by use of a fixed-effects or a random-effects model, which are the subjects of this chapter.

6.1.1 Fixed- or Random-Effects Models

There is no unanimity among statisticians as to what constitutes a fixed-versus a random-effects model. The most common distinction is that a fixed-effects model assumes that the observations in a panel all come from a fixed source. The differences in the observations within panels are therefore not important. In addition, a fixed-effects model assumes that the primary interest of the study are the measurements themselves.

A random-effects model, on the other hand, assumes that the measure-ments within panels are representative of a greater population. Inference to a greater population is therefore of paramount importance. Typically the ran-dom effects are given as normally distributed with a mean of 0 and variance of σ^2, but this does not need to be the case.

We shall discuss each type of model in the following two sections. Our primary interest, however, is not in the theory of panel models, but rather in the code that can be used to estimate panel models in general. An overview of the logic of these two types of models will be given together with a brief look at some of their characteristics and caveats.

6.2 Fixed-Effects Model

6.2.1 Unconditional Fixed-Effects Models

A fixed-effects model may be parameterized as unconditional or as conditional. Unconditional fixed-effects estimators include a dummy or indicator variable for each of the $J - 1$ panels in the data. The same model that was used for non-panel or pooled data is used to estimate the unconditional model. In R this is done by use of the **factor** option, where the individual panels in the data are each given an intercept, with the lowest level being the reference. The model can be rather cumbersome if there are more than a few panels in the model.

For example, we model the **medpar** data as we did above, but add a factor predictor for hospital (**provnum**). Since there are 54 distinct hospitals, the model will display 53 separate coefficients that are related to the reference level, *provnum* = 030001, in addition to the parameter estimates of **hmo** and **white**. We also recode the factor to save space.

```
> medpar$pr <- factor(substr(medpar$provnum, 3, 6))
> ufenb <- ml_glm2(los ~ hmo + white + pr,
+                  formula2 = ~1,
+                  data = medpar,
+                  family = "negBinomial",
+                  mean.link = "log",
+                  scale.link = "identity_s")
```

The reader may wish to compare the speed of convergence of **ml_glm2** with that of (the superior) **glm.nb**. To save space we display only the first and last few fixed panel effects of the model output.

```
> summary(ufenb)

Call:
ml_glm2(formula1 = los ~ hmo + white + pr, formula2 = ~1,
    data = medpar, family = "negBinomial", mean.link = "log",
    scale.link = "identity_s")

Deviance Residuals:
    Min. 1st Qu.  Median    Mean 3rd Qu.    Max.
 -2.8990 -0.8706 -0.2023 -0.2114  0.4889  3.6050
```

Coefficients:

	Estimate	SE	Z	p	LCL	UCL
(Intercept)	1.9851	0.1154	17.2044	2.46e-66	1.75896	2.2113
hmo	-0.1063	0.0548	-1.9400	0.0524	-0.21364	0.0011
white	-0.0177	0.0713	-0.2484	0.804	-0.15746	0.1220
...						
pr0093	0.2861	0.1465	1.9528	0.0508	-0.00105	0.5732
pr0094	0.1565	0.2362	0.6625	0.508	-0.30648	0.6195
pr2000	1.3115	0.1437	9.1291	6.91e-20	1.02991	1.5931
pr2002	1.3724	0.2279	6.0206	1.74e-09	0.92559	1.8191

```
pr2003          1.8933 0.4639  4.0812  4.48e-05  0.98405  2.8026
(Intercept)_s   0.3905 0.0181 21.5281  8.5e-103  0.35499  0.4261
```

```
Null deviance: 1868.147  on  1493 d.f.
Residual deviance: 1565.204  on  1438 d.f.
AIC:  9551.933
```

```
> nbtest3 <- glm.nb(los ~ hmo + white + pr, data = medpar)
> P__disp(nbtest3)              # Pearson dispersion
```

```
pearson.chi2   dispersion
 1492.374547    1.037091
```

The dispersion statistic has reduced to 1.037 from 1.329. Clearly the unconditional fixed-effects model is superior to the standard negative binomial model. It is also evident that modelling all 54 hospitals this way is messy, with many **provnum** levels not significantly contributing to the model. Probably many levels can be combined. As a rule, contiguous panels may be combined if their slopes are near identical, or if one panel is a significant level and one or more contiguous levels to it are not. If the level contiguous to the reference level is not significant, one may refrain from entering it in the model. In such a case the new reference is the combined first two levels of the factored panel variable. Note that unless one actually encodes the first two levels to form a single enlarged level, each non-reference level must be specifically indicated in the model. For any of these modelling options, it is necessary to test the new relationships, for example using a likelihood ratio test.

An unconditional fixed-effects model may be appropriate if it comprises no more than four to ten or twelve levels. Such models are also preferred when one wishes to know the values of the various fixed panel-level parameters in the model. When there are a large number of levels, most statisticians will select a conditional fixed-effects model if the data consists of fixed panels.

6.2.2 Conditional Fixed-Effects Models

Unconditional fixed-effects models are estimated using standard GLM software. Conditional fixed-effects models, however, are derived from a different parameterization of the underlying exponential family of distributions from which generalized linear models are based. That is, conditional fixed-effects models amend the GLM PDF and log-likelihood to accommodate the panel nature of the data. An extra parameter, γ, is provided to the model. γ symbolizes the unknown cluster-specific parameters, which are independent across panels.

$$y_{ij} = X_{ij}\beta_j + \gamma_j + \epsilon_{ij} \tag{6.1}$$

The key to understanding conditional fixed-effects models is to under-

stand that the fixed-panel effects are conditioned out of the likelihood function through the sufficient statistic y_{ij}. In the case of the negative binomial,
the heterogeneity parameter is also conditioned out, leaving a log-likelihood
that appears as follows.

$$
\begin{aligned}
\mathcal{L}(\lambda|y) = \sum_{i=1}^{n} & \left[\log \Gamma \left(\sum_{t=1}^{n_i} \lambda_{it} \right) + \log \Gamma \left(\sum_{t=1}^{n_i} y_{it} + 1 \right) \right. \\
& - \log \Gamma \left(\sum_{t=1}^{n_i} \lambda_{it} + y_{it} \right) \\
& \left. + \sum_{t=1}^{n_i} \left[\log \Gamma(\lambda_{it} + y_{it}) - \log \Gamma(\lambda_{it}) - \log \Gamma(y_{it}) \right] \right]
\end{aligned}
\tag{6.2}
$$

Note that neither γ nor the negative binomial heterogeneity parameter α
is included in the log-likelihood. Except for the many summations employed
in the function, the log-likelihood is relatively simple to estimate.

Also note that the conditional fixed-effects negative binomial model is
based on the linear parameterization of the negative binomial, commonly referred to as NB1 (Hilbe, 2011). The NB1 likelihood is

$$
L(\lambda|y,\delta) = \frac{\Gamma(y_i + \lambda_i)}{\Gamma(y_i + 1)\Gamma(\lambda_i)} \left(\frac{\delta}{1+\delta} \right)^{\lambda_i} \left(\frac{1}{1+\delta} \right)^{y_i}
\tag{6.3}
$$

which includes the heterogeneity parameter, δ. Conditioning on the sufficient
statistic eliminates δ. The relationship of NB1 and the conditional fixed-effects
model is fully discussed in Hardin and Hilbe (2012). δ is understood the same
way as α, the heterogeneity parameter of the traditional parameterization of
negative binomial, NB2.

The difference in the NB1 and NB2 derives from their respective variance
functions. With μ symbolizing the same mean parameter as λ, the NB1 and
NB2 models have variance functions defined as $\mu + \alpha\mu$ and $\mu + \alpha\mu^2$, respectively. The '1' and '2' associated with NB in the above equations indicate the
degree of the variance function, with NB1 being linear and NB2 quadratic.
The symbol μ is nearly always used with respect to generalized linear models,
whereas λ is generally used to indicate the mean parameter for count models
that are estimated using a non-IRLS full maximum likelihood routine.

A foremost concern of panel models in general is the relationship of
between-panel variation to within-panel variation. When panels deal with observations on people over time, the consideration is of the relationship of
between-person to within-person variation. For fixed characteristics such as
gender, race, date-of-birth and so forth, there is only between-person variation. A specific person will be male, for example, across other characteristics,
with no change in value (unless the study relates to sex change operations
for instance). For our example, panels are of hospitals, which do not change

within panels. Conditional fixed-effects models, by eliminating fixed panels or items from the model, focus on explaining within-panel variation for items or characteristics that are not fixed. Between-item or person variation, however, is not estimated by the conditional fixed-effects model. The result of this is an inflation of standard errors. Random-effects models solve this problem though by modelling both within and between item variation, and by having γ follow a specific probability distribution whose parameters are estimated by the model (Hilbe, 2009). We address random-effects models in the next section.

It should be mentioned that each type of conditional fixed-effects model has features that may not be shared by other conditional fixed-effects models. For instance, the negative binomial conditional fixed-effects model has an intercept, whereas the corresponding Poisson model does not. Excellent discussions of these model characteristics can be found in Hsiao (2003) and Frees (2004).

We now address an approach that can be taken to estimate conditional fixed-effects negative binomial parameter estimates. We do not estimate a value of the heterogeneity parameter, nor of the fixed-effects parameter. The key to the solution is accounting for the panel structure of the data, which must be summed within panels and then across panels. We note that R has not previously supported this model to our knowledge. We have compared its estimates, standard errors and associate statistics with Stata output on the same data.

6.2.3 Coding a Conditional Fixed-Effects Negative Binomial

We selected to encode conditional fixed-effects negative binomial because it can be estimated using maximum likelihood methodology and since neither the heterogeneity parameter, δ, nor the parameter for the within-item panel effect needs to be estimated. Not all fixed-effects models are as simple, and not all can be estimated using maximum likelihood, requiring techniques such as EM and quadrature. This will be particularly the case when dealing with random-effects models.

We may adapt the previous code used for estimating a model using maximum likelihood for the conditional fixed-effects negative binomial. Without having to estimate any parameter other than the mean, which we may symbolize as either μ or λ, estimation is relatively simple. The challenge is in summing observations within samples, then summing across panels. It is also easier to break up the log-likelihood function into separate terms, performing the required operations separately on each.

We program the conditional fixed-effects negative binomial log-likelihood as:

```
> jll.gNegBinomial <- function(y, y.hat, groups, ...) {
+   y.hat.sum.g <- tapply(y.hat, groups, sum)
+   y.sum.g <- tapply(y, groups, sum)
```

```
+    both.sum.g <- y.hat.sum.g + y.sum.g
+    lg.y.hat.sum.g <- tapply(lgamma(y.hat), groups, sum)
+    lg.y.sum.g <- tapply(lgamma(y + 1), groups, sum)
+    lg.both.sum.g <- tapply(lgamma(y + y.hat), groups, sum)
+    ll <- sum(lgamma(y.hat.sum.g) + lgamma(y.sum.g + 1) -
+             lgamma(both.sum.g) + lg.both.sum.g -
+             lg.y.sum.g - lg.y.hat.sum.g)
+ }
```

We have made liberal use of the excellent `tapply` function, which allows for the calling of a function across levels of a factor, here the grouping structure.

Note that this likelihood is a function of the linear predictor and the data. We can re-use much of the programming infrastructure that was developed in the first part of Chapter 4, which we do not list here again. The reader should ensure that the objects defined in that chapter are loaded, ideally by loading the *msme* package.

```
> library(msme)
```

We need to do one more thing. It turns out that fitting a null model is problematic with the setup we developed in Chapter 5. So, we rewrite our `ml_glm2` function to drop the null model. We refer to this version as `ml_glm3`.

```
> ml_glm3 <- function(formula,
+                      data,
+                      family,
+                      link,
+                      offset = 0,
+                      start = NULL,
+                      verbose = FALSE,
+                      ...) {
+    mf <- model.frame(formula, data)
+    y <- model.response(mf, "numeric")
+    class(y) <- c(family, link, "expFamily")
+    X <- model.matrix(formula, data = data)
+    if (any(is.na(cbind(y, X)))) stop("Some data are missing!")
+    if (is.null(start))  start <- kickStart(y, X, offset)
+    fit <- maximize(start, Sjll, X, y, offset, ...)
+    if (verbose | fit$convergence > 0)  print(fit)
+    beta.hat <- fit$par
+    se.beta.hat <- sqrt(diag(solve(-fit$hessian)))
+    residuals <- devianceResiduals(y, beta.hat, X, offset, ...)
+    results <- list(fit = fit,
+                    X = X,
+                    y = y,
+                    call = match.call(),
```

```
+                    obs = length(y),
+                    df.null = length(y) - 1,
+                    df.residual = length(y) - length(beta.hat),
+                    deviance = sum(residuals^2),
+                    null.deviance = NA,
+                    residuals = residuals,
+                    coefficients = beta.hat,
+                    se.beta.hat = se.beta.hat,
+                    aic = - 2 * fit$val + 2 * length(beta.hat),
+                    i = fit$counts[1])
+    class(results) <- c("msme","glm")
+    return(results)
+ }

> data(medpar)
```

The code to fit the conditional fixed-effects negative binomial model is now
run using the familiar function call.

```
> med.nb.g <- ml_glm3(los ~ hmo + white,
+                      family = "gNegBinomial",
+                      link = "log",
+                      group = medpar$provnum,
+                      data = medpar)
```

We can obtain the usual important summary information from our function
as follows.

```
> summary(med.nb.g)

Call:
ml_glm3(formula = los ~ hmo + white, data = medpar,
    family = "gNegBinomial", link = "log",
    group = medpar$provnum)

Deviance Residuals:
   Min. 1st Qu. Median   Mean 3rd Qu.   Max.
 -51.14   51.14  51.14  37.66   51.14  51.14


Coefficients:
            Estimate      SE      Z         p    LCL      UCL
(Intercept)  0.92927 0.07581 12.258 1.525e-34 0.7807 1.07786
hmo         -0.07014 0.05153 -1.361    0.1735 -0.1711 0.03086
white       -0.02698 0.06533 -0.413    0.6796 -0.1550 0.10106

Null deviance: NA  on  1494 d.f.
```

```
Residual deviance: 3909331   on   1492 d.f.
AIC:   8992.173
```

```
Number of iterations:   248
```

We checked these results against the same model fit using Stata's xtnbreg function, and the estimates, standard errors, and fit statistics are identical.

We see that neither hmo nor white significantly contribute to the model fit. Neither did they contribute to the unconditional model. Recall that both hmo and white appeared to be significant when used in a traditional negative binomial model, but the fact that the model was over-dispersed meant that the estimates of the standard errors were biased.

It should be noted that the negative binomial conditional-effects model is not a true fixed-effects model. Allison and Waterman (2002) discovered that the model fails to control for all of its predictors. Several statisticians attempted to resolve the problem by using a negative multinomial model in place of a conditional fixed-effects negative binomial, but it was discovered that the estimates were identical to the conditional Poisson, and therefore do not appropriately accommodate any excess dispersion over that modeled by the conditional Poisson (Hilbe, 2011). Allison and Waterman (2002) recommend using the unconditional fixed-effects model rather than the conditional, but in cases where there are a host of panels such a recommendation is not satisfactory. We recommend that an unconditional fixed-effects model be used in place of an conditional model if possible. For data situations in which there are a large number of panels it may be preferred to employ a random-effects or a generalized estimating equation model.

6.3   Random-Intercept Model

6.3.1   Random-Effects Models

Random-effects models are appropriate when the observations being measured or recorded are assumed to be randomly drawn from a greater population that is distributed according to some known probability distribution. This set of models is not intended to directly estimate random effects, but rather a variance component from the distribution of the random effect. An important assumption that is characteristic of random-effects models is that the random effects are uncorrelated with the explanatory predictors.

Modelling an effect as random usually — although not necessarily — assumes that the random effects are normally distributed. Often, however, this is not the case. For example, with random-effects count models the random-effects term can be assumed to follow the gamma distribution. Whether one assumes a non-normal distribution for the random effect is usually driven by whether the resulting joint distribution has an analytic solution. As such, we focus on normal effects to keep the model as elementary as possible.

Many researchers use the term *random effects* referring to what is now commonly called a random-intercept model. The random-intercept model is a subset of random-coefficient models. The larger set of models allow model coefficients to vary between panels; the random-intercept model allows only the intercept to vary between panels. For a longitudinal study where each panel in the data consists of measurements made on an individual over time, a random-intercept model of the data comprises a separate intercept for each panel or individual. We shall examine a Gaussian random-intercept model where panels, or individuals, are normally distributed with a zero mean and estimated random-intercept variance σ^2.

Random-intercept models have been symbolized in a variety of ways. The following is a common way of symbolizing the basic random-intercept model:

$$y_{it} = \alpha_i + \mathbf{X}_{it}\beta + \epsilon_{it} \qquad (6.4)$$

with α_i indicating the random intercept for each panel in the model, $\mathbf{X}_{it}$ representing data structured with observations, i, within panels, t. β is the coefficient or slope of the predictors, and ϵ_{it} are the observation-level model errors measured for each individual within panels. The unobserved random intercepts are assumed to follow a normal distribution with mean 0 and variance σ_α^2 and are sometimes referred to as *random effects*. It should be noted that we used γ as the parameter for the panel effect in the fixed-effects model in the previous section. α is used here to clearly differentiate between the two types of models. Each is dealt with by their respective model in entirely different ways.

The random effects may be correlated within panels but are independent between panels. A random-intercept model is best estimated if the model consists of ten or more panels each with two or more observations. Random-effects models may include single-unit panels, but such panels obviously provide no information regarding within-panel correlation, or information regarding the relationship of within-panel to between-panel variance. Single panels do, however, contribute to estimation of the model βs.

Typical output from a Gaussian random intercept model includes statistics for both the standard deviation of α_i, the random intercepts, and of the standard deviation of ϵ_{it}, the residuals. If the standard deviation of α_i significantly differs from zero, the intercepts do vary, and contribute to the extra correlation in the data. A likelihood ratio test may also be used to evaluate if the data are best modeled using a random-effects model or a pooled model; i.e., a model where panels are pooled and have no effect on the model. Both statistics often thought to be distributed as Chi2 with one degree of freedom; however, see Stram and Lee (1994), Pinheiro and Bates (2000), and our Section 7.2.3. p-values under 0.05 indicate that the random-intercept model is preferred.

Keep in mind that there are more complex random-effects models, as well as mixed-effects models. We have described the most elementary model of this set of models where only the intercept varies across panels. However, it is a good model to use for explaining the logic of this set of models in general. Also, note that random-effects models, including random intercept models, are many times estimated by means other than maximum likelihood. We show how to encode a maximum-likelihood Gaussian random-intercept model in this chapter, and will leave discussion of the EM algorithm to the following section.

We shall use the `ufc` data from the *msme* package for an example of a panel model that is suitable for estimation using a Gaussian random-intercept model. The data relates to a forestry study. We propose to construct a model with tree height (`height.m`) as the response and tree bole (trunk) diameter at 1.37 m from the ground (`dbh.dm`) as the predictor. Panels consist of plots (`plot`), which can be regarded as a representative sample from the greater population of all plots. There are 144 distinct plots with a total of 637 observations, only about half of which have measures for both variables. Panels of plots consist of 1 to 13 observations. DBH size ranges from 100 to 1120 cm. The response variable, `height.m`, ranges from 0 (1 observation) to 380 cm.

Before describing code that can be used to estimate a basic random-effects model, we load and display the type of data that will be used in the model.

```
> library(msme)
> data(ufc)

> ufc <- na.omit(ufc)
> head(ufc)

   plot tree species dbh.cm height.m
2     2    1      DF     39     20.5
3     2    2      WL     48     33.0
5     3    2      GF     52     30.0
8     3    5      WC     36     20.7
11    3    8      WC     38     22.5
12    4    1      WC     46     18.0
```

The basic descriptions of the response variable, height.m, and predictor, dbh.cm, are displayed as:

```
> summary(ufc$height.m)

  Min. 1st Qu.  Median    Mean 3rd Qu.    Max.
  0.00   18.85   24.00   24.07   29.15   48.00

> summary(ufc$dbh.cm)

  Min. 1st Qu.  Median    Mean 3rd Qu.    Max.
 10.00   22.85   33.50   36.44   46.30  112.00
```

The *nlme*, *gamlss.mx*, *lme4*, and *plm* packages can all be used to estimate a wide range of fixed-, random-, and mixed-effects models. All four may be downloaded from CRAN, and the first is included in the base R installation. Here we use the `lme` function from the *nlme* package (Pinheiro and Bates, 2000).

```
> library(nlme)
> renorm.lme <- lme(height.m ~ dbh.cm,
+                   random = ~ 1 | plot,
+                   data = ufc, method = "ML")
> summary(renorm.lme)

Linear mixed-effects model fit by maximum likelihood
 Data: ufc
       AIC      BIC    logLik
  2362.745 2378.62 -1177.373

Random effects:
 Formula: ~1 | plot
        (Intercept) Residual
StdDev:    1.279798 4.758336

Fixed effects: height.m ~ dbh.cm
                Value Std.Error  DF  t-value p-value
(Intercept) 12.698399 0.5760215 256 22.04501       0
dbh.cm       0.310401 0.0140255 256 22.13119       0
 Correlation:
        (Intr)
dbh.cm -0.884

Standardized Within-Group Residuals:
        Min          Q1         Med          Q3         Max
-6.78362499 -0.55246700  0.02914238  0.53189967  2.50511273

Number of Observations: 391
Number of Groups: 134
```

6.3.2   Coding a Random-Intercept Gaussian Model

We first must provide the log-likelihood function for the random-intercept Gaussian model. With the variance of the random intercepts symbolized as σ_u^2 and the variance of the errors as σ_ϵ^2, a standard formulation for the log-likelihood of the i-th panel is

$$\mathcal{L} = -\frac{1}{2} \frac{\sum_t z_{it}^2 - \gamma_i (\sum_t z_{it})^2}{\sigma_\epsilon^2} + \ln\left(n_i \frac{\sigma_u^2}{\sigma_\epsilon^2} + 1\right) + n_i \ln(2\pi\sigma_\epsilon^2) \qquad (6.5)$$

where i is the panel indicator, t the within-panel indicator, n_i is number of observations in i-th panel, $z_{it} = y_{it} - x_{it}\beta$, and

$$\gamma_i = \frac{\sigma_u^2}{n_i \sigma_u^2 + \sigma_\epsilon^2} \tag{6.6}$$

A pseudo-code setup of the log-likelihood function simplifies understanding complex models. Pseudo-code is code that appears close to the code that will be used to estimate the model. It provides the logic of the log-likelihood function. Note that we use `Su` to indicate the natural log of the standard deviation of the intercepts and `Se` to symbolize the natural log of the standard deviation of the errors, that is, we will use the log link function for the variance components. It is necessary to use the standard deviations here because terms will be used to modify them prior to squaring.

$$g = \exp(Se)^2/(N * \exp(Su)^2 + \exp(Se)^2)$$
$$a = \left(\sum_t (z^2)\right) - g \times \sum_t (z)^2/\exp(Se)^2$$
$$b = \ln(N * \exp(Su)^2/\exp(Se)^2 + 1)$$
$$c = N * \ln(2 * \pi * \exp(Se)^2)$$
$$\mathcal{L} = \sum(-0.5 * (a + b + c))$$

Translating this pseudo-code into R gives the following code as a definition of the objective function.

```
> jll_gnormal <- function(params, y, X, group, ...) {
+     p <- ncol(X)
+     N_i <- tapply(y, group, length)
+     Su <- exp(params[p+1])
+     Se <- exp(params[p+2])
+     z <- y - X %*% params[1:p]
+     gamma_i <- Su^2 / (N_i * Su^2 + Se^2)
+     c1 <- (tapply(z^2, group, sum) -
+             gamma_i * tapply(z, group, sum)^2) / Se^2
+     c2 <- log(N_i * Su^2 / Se^2 + 1)
+     c3 <- N_i * log(2 * pi * Se^2)
+     return(sum(-0.5 * (c1 + c2 + c3)))
+ }
```

We now gather the pieces of data that we need to populate and then maximize this function. For convenience, we will place them in a list object. Following the examples above, y will be tree height, X will be a design matrix constructed for the regression of tree height upon tree diameter, and the group structure will be defined by the forest plot.

```
> ufc.model <- with(ufc,
+                    list(y = height.m,
+                            X = model.matrix(~ dbh.cm),
+                            group = plot))
```

We pick some arbitrary starting points for our parameter estimates to provide to the optimizer, and try to evaluate the objective function at those points.

```
> start <- c(1,1, 1, 1)
> with(ufc.model, jll_gnormal(start, y, X, group))
```

```
[1] -5286.406
```

This result seems satisfactory. We can now call **optim** on the objective function using these data and start points with some confidence.

```
> ufc.fit <-
+    with(ufc.model,
+          optim(start,
+                 jll_gnormal,
+                 y = y,
+                 X = X,
+                 group = group,
+                 method = "BFGS",
+                 control = list(fnscale = -1,
+                                  reltol = .Machine$double.eps,
+                                  maxit = 10000)))
```

The fitting exercise has converged. The astute reader will note that we have used the **method** and **control** arguments to change the nature and behavior of the optimizer. The same reader may wish to experiment with the default settings and these data and model.

The parameter estimates for the fixed and random effects, as fitted above, are the same as obtained in the earlier demonstration, to three significant digits. First we compare the fixed effects, with the *lme* object presented first:

```
> fixed.effects(renorm.lme)
```

```
(Intercept)        dbh.cm
 12.6983992    0.3104011
```

```
> ufc.fit$par[1:2]
```

```
[1] 12.6983993  0.3104011
```

then the random effects, *lme* object presented first:

```
> VarCorr(renorm.lme)[,2]
```

```
(Intercept)      Residual
 "1.279798"   "4.758336"

> exp(ufc.fit$par[3:4])

[1] 1.279796 4.758338
```

6.4   Handling More Advanced Models

We indicated that the random-intercept model is a subset of the more general random-coefficient model, for which the coefficients of the slopes are also allowed to vary across panels. This type of model is nearly always estimated using an EM algorithm, or by either adaptive or Gaussian quadrature, penalized least squares, generalized least squares, or by simulation. In order to capture the effects of both slopes and intercepts varying the general random-effects model is generally symbolized as:

$$y_{it} = \mathbf{X}_{it}\beta_t + b_t\mathbf{Z}_{it} + \epsilon_{it} \tag{6.7}$$

where ζ_t are the random coefficients of the model, including the intercept, and z are the predictors or covariates corresponding to the random effects. $\mathbf{X}\beta$ are the fixed effects. A random-intercept model will be indicated as having the random component symbolized as $b_t Z_{0t}$. This formulation leads to the more general set of mixed and multilevel models.

We have been discussing two-level models in the chapter. Three and four-level nested models can also be developed by adding more terms to Equation 6.7, an extra term for each level. Partitioning more levels of variability does not come without problems though. We recommend Gelman and Hill (2007) for a thorough discussion of these types of models. The authors use R for all examples.

6.5   The EM Algorithm

The EM algorithm is an iterative method used to determine the maximum likelihood estimator of a model parameter when some of the data in the model are unobserved. The method was first defined and referred to as EM by Dempster et al. (1977), with their acknowledgment that forms of the method had previously been used in research by others. For example, Hartley (1958) used a technique very similar to EM when dealing with multinomial models having missing data. He admits in his article, though, that the method he used was old and varied. Dempster et al. trace EM back to McKendrick (1926), and Efron in his discussion of Dempster et al.'s paper mentions Fisher (1925) as having anticipated the method. It should also be noted that the Baum–Welch algorithm presented by Baum et al. (1970) for modelling hidden Markov models in speech recognition used the same method as now recognized as EM. The use of EM in related fields is largely due to this paper. We recommend Meng and van Dyk (1997) for an excellent overview of the method and its origin. McLachlan and Krishnan (2008) is recommended as an excellent text on the method and its variations.

Missing and unobserved data have been a problem in statistics for a long time. The EM algorithm is now primarily used to handle models with missing values, as well as models with random effects, latent and censored variables, and models with other types of multilevel and panel structures. The method generally used for missing values, however, is a bit different from how the method is employed when finding maximum likelihood estimates of unobserved parameters. In fact, there are a host of variants of the basic EM method. The acronym EM represents "Expectation–Maximization," which indicates that it is a two-stage process.

The first stage of the method separates the data into two components — (1) where data or parameters are observed, and (2) for data that is missing or parameters that are unobserved. The first step calculates a full maximum likelihood estimation on the observed data and parameters. The conditional expectation of the missing data or parameters are obtained. This is the E-step.

The next stage is to maximize the data using the conditionally expected values in the place of missing values or unobserved parameters. This is the M-step.

The conditional expected values for the original missing values are again calculated, and are inserted as new values into the data. This is the second stage E-step. A full MLE of the data including the imputed values is obtained — the second stage M-step. The procedure continues to when there is very little change in the parameter estimates of the M-step model.

The EM algorithm has a number of statistically appealing properties, including:

1. It can be used to provide estimates of missing values.

2. Most of the leading software applications support EM.

3. It can produce unbiased parameter estimates for models with missing values, and with adjustment, unbiased standard errors.

Likewise, there may be complications with implementing the method, including:

1. It can be slow, taking hundreds or more iterations to converge.

2. The variance-covariance matrix is not estimated; however, with adjustments it can usually be determined.

3. The algorithm can be difficult to implement for various modelling situations.

It is also important to keep in mind that when using EM for imputing missing values, it is assumed that the missing values are either missing completely at random (MCAR) or at least missing at random (MAR). If there is a correlation to the missingness, or missing values come in blocks, then EM methodology will not be appropriate. When missing values are MCAR, they are randomly distributed across all observations in the data. When MAR, the missing values are randomly distributed within one of more subsamples of the data. See Hardin and Hilbe (2002) for a discussion of missingness in general and its application to both pooled and panel models.

6.5.1   A Simple Example

We demonstrate a simple example of an EM algorithm for imputing missing values using the `ufc` data. Note that we eliminated the missing values during our previous usage, so we need to reload the data.

```
> library(msme)
> data(ufc)
```

We next examine the missingness pattern with the `ufc` dataframe. We do this by using the `sapply` function upon a bespoke function that simply sums the count of missing values on a vector.

```
> sapply(ufc, function(x) sum(is.na(x)))

   plot      tree   species    dbh.cm  height.m
      0         0         0        10       246
```

We then eliminate the responses that correspond to empty plots, namely those that have missing values for `dbh.cm`.

```
> ufc <- ufc[!is.na(ufc$dbh.cm),]
```

We now record which observations have missing height values in a variable called `na`, and replace the missing heights with the mean height.

```
> ufc$na <- is.na(ufc$height.m)
> ufc$height.m[ufc$na] <- mean(ufc$height.m, na.rm = TRUE)
```

We are in position to begin looping. Here we just use a tuned `for` loop; we could also apply a `while` function, but this is left as an exercise for the reader. The loop comprises three steps, two major and one minor: we fit the model to the complete dataset (this is the *Maximization* step), then use the model to predict the missing heights (this is the *Expectation* step), and finally we record the coefficients for post-hoc reporting.

```
> reps <- 20
> trace <- vector(mode = "list", length = reps)
> for (i in 1:reps)  {
+    refit.lm <- lm(height.m ~ dbh.cm,
+                   data = ufc,
+                   na.action = na.exclude)         # M step
+    ufc$height.m[ufc$na] <-
+      predict(refit.lm, newdata = ufc)[ufc$na]     # E step
+    trace[[i]] <- coef(refit.lm)
+ }
```

The combination of functions `do.call` and `rbind` provide a convenient way to stitch the elements of the `trace` list together.

```
> trace <- do.call(rbind, trace)
```

Figure 6.1 shows the trajectory of the EM parameter estimates, along with a cross at the least-squares estimates, which is in any case the location of convergence. It is constructed using the following code.

```
> par(las = 1, mar=c(4,4,2,1))
> plot(dbh.cm ~ '(Intercept)', data = trace, type = "b")
> base.model <- lm(height.m ~ dbh.cm, data = ufc)
> points(coef(base.model)[1], coef(base.model)[2],
+        cex = 2, pch = 3)
```

6.5.2 The Random-Intercept Model

We now demonstrate how to fit the random-intercept model using the EM algorithm. This example is instructive because it shows that despite the conceptual simplicity of EM, quite a lot of work has to be done to make it operational in real settings.

First, we review the algorithm, which is from Demidenko (2004). Let the model for the data in the i-th group be

$$y_i = \mathbf{X}_i\beta + \mathbf{Z}_i b_i + \epsilon_i, \quad i = 1, \ldots, n. \tag{6.8}$$

where the definitions are as previously. Let $\epsilon_i \sim N(0, \sigma^2)$ and $b_i \sim N(0, \sigma^2 \mathbf{D})$. Let N_T be the count of all the observations. Denote

$$V_i = \mathbf{I} + \mathbf{Z}_i \mathbf{D} \mathbf{Z}_i' \tag{6.9}$$

The updating algorithms to go from s to $s+1$ are then

$$\sigma_{s+1}^2 = \sigma_s^2 - 1 + \frac{1}{\sigma_s^2 N_T} \sum_{i=1}^{N} (\mathbf{y}_i - \mathbf{X}_i\beta_s)' \mathbf{V}_{is}^{-1} (\mathbf{y}_i - \mathbf{X}_i\beta_s) \tag{6.10}$$

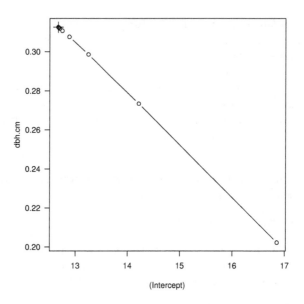

FIGURE 6.1
Trajectory of EM-based parameter estimates for the regression of tree height upon diameter for the `ufc` dataset.

$$\mathbf{D}_{s+1} = \mathbf{D}_s - \frac{1}{N} \sum_{i=1}^{N} \left[\mathbf{D}_s \mathbf{Z}_i' \mathbf{V}_{is}^{-1} \mathbf{Z}_i \mathbf{D}_s \right.$$
$$\left. - \mathbf{D}_s \mathbf{Z}_i' \mathbf{V}_{is}^{-1} \mathbf{e}_{is} \mathbf{e}_{is}' \mathbf{V}_{is}^{-1} \mathbf{Z}_i \mathbf{D}_s / \sigma_s^2 \right] \tag{6.11}$$

$$\beta_{s+1} = \left[\sum_{i=1}^{N} \mathbf{X}_i' (\mathbf{I} + \mathbf{Z}_i \mathbf{D}_s \mathbf{Z}_i')^{-1} \mathbf{X}_i \right]^{-1} \sum_{i=1}^{N} \mathbf{X}_i' (\mathbf{I} + \mathbf{Z}_i \mathbf{D}_s \mathbf{Z}_i')^{-1} \mathbf{y}_i \tag{6.12}$$

We code the algorithms in the following way. The representation is group-by-group, which keeps the memory overhead comparatively low. In order to retain this structure of the algorithm conveniently, we create lists of matrices, and use the `mapply` function to manipulate them elementwise. First, we recover the data, and strip out the missing values.

```
> data(ufc)
> ufc <- na.omit(ufc)
```

The following declarations are definitions from the model specification.

```
> (N_T <- nrow(ufc))
> N <- length(unique(ufc$plot))
> y_i <- with(ufc, split(height.m, plot))
> x_i <- with(ufc, split(dbh.cm, plot))
> X_i <- lapply(x_i, function(x) cbind(1, x))
> Z_i <- lapply(x_i, function(x) matrix(1, nrow = length(x)))
```

We need to find suitable start points. We take the residual variance from an OLS fit to the data as σ_0^2, the variance of the plot-level means of the residuals as $\mathbf{D}_0$, and the coefficients from the same model as the starting parameter estimates β_0.

```
> lm.start <- with(ufc, lm(height.m ~ dbh.cm))
> (s2 <- summary(lm.start)$sigma^2)
```

```
[1] 24.41567
```

```
> (D <- with(ufc,
+               var(tapply(residuals(lm.start),
+                          plot,
+                          mean))))
```

```
[1] 11.84213
```

```
> beta <- coef(lm.start)
> dim(beta) <- c(length(beta), 1)
```

We also need to define the inverse of the variance matrix $\mathbf{V}$. Here we do so as elements of a list.

```
> V_inv <- lapply(Z_i,
+                 function(Z, D)
+                 solve(diag(nrow(Z)) + Z %*% D %*% t(Z)),
+                 D = D)
```

We can now begin the loop. In this variation we update the estimates sequentially because the program flow is easier to follow. Hence the code below does not quite marry with Equations 6.9–6.12.

```
> for (i in 1:2000) {
+
+ ## First, compute the list of E_i as within-group residuals
+ ## from the fixed effects.
+
+   E_i <- mapply(function(y, X, beta)
+                 y - X %*% beta,
+                 y = y_i, X = X_i, beta = list(beta))
```

```
+
+ ## Then obtain the within-group variance contribution by the
+ ## weighted inner product of the residuals for each group.
+
+   s2.i <- mapply(function(E, V)
+                  sum(t(E) %*% V %*% E),
+                  E = E_i, V = V_inv)
+
+ ## Sum these across the groups, scale them, and use them to
+ ## update the variance estimate.
+
+   s2 <- s2 - 1 + sum(s2.i) / (s2 * N_T) # This is 6.10
+
+ ## D_i is the groupwise calculated correction to matrix D.
+ ## Refer to Equation 6.11; the following is the quantity
+ ## within the squared brackets.
+
+   D.i <-
+     mapply(function(D, Z, V, E, s2)
+            D %*% t(Z) %*% V %*% Z %*% D -
+            D %*% t(Z) %*% V %*% E %*% t(E) %*%
+             V %*% Z %*% D / s2,
+             D = list(D), Z = Z_i, V = V_inv, E = E_i,
+             s2 = list(s2))
+
+   D <- D - sum(D.i) / N                   # This is 6.11
+
+ ## We can now update V, given the new D.  Note our use of
+ ## diag(nrow(Z)) to create a suitably sized identity matrix
+ ## on the fly.
+
+   V_inv <- lapply(Z_i,
+                   function(Z, D)
+                   solve(diag(nrow(Z)) + Z %*% D %*% t(Z)),
+                   D = D)                   # This is 6.9
+
+ ## Finally we update Beta
+
+   beta.denom <-
+     mapply(function(X, D, Z)
+            t(X) %*% solve(diag(nrow(Z)) +
+             Z %*% D %*% t(Z)) %*% X,
+             X = X_i, D = list(D), Z = Z_i, SIMPLIFY = FALSE)
+
+   beta.num <-
```

```
+       mapply(function(X, D, Z, y)
+               t(X) %*% solve(diag(nrow(Z)) +
+               Z %*% D %*% t(Z)) %*% y,
+               X = X_i, D = list(D), Z = Z_i, y = y_i,
+               SIMPLIFY = FALSE)
+
+     beta <- solve(Reduce('+', beta.denom)) %*%
+     Reduce('+', beta.num)                     # 6.12
+
+ }
```

The only new function used here is `Reduce`, which we use to sum the lists of matrices, elementwise. See `?Reduce` for its general purpose. After a sufficiently large number of iterations, we examine the parameter estimates. The EM estimates of the fixed effects are here compared with the result from `lme`, with the latter presented first.

```
> fixed.effects(renorm.lme)
```

```
(Intercept)        dbh.cm
 12.6983992     0.3104011
```

```
> beta[,1]
```

```
                    x
12.6983992   0.3104011
```

Next we compare the estimated random effects from each model, again with that from `lme` first.

```
> VarCorr(renorm.lme)[,1]
```

```
(Intercept)      Residual
" 1.637883"   "22.641760"
```

```
> c(D*s2, s2)
```

```
[1]   1.637883 22.641760
```

All of the EM parameter estimates compare well with those from `lme`.

6.6   Further Reading

There are numerous excellent texts on the analysis of panel models. We particularly admire Pinheiro and Bates (2000), Schabenberger and Pierce (2002), Fitzmaurice et al. (2004), Demidenko (2004), and Gelman and Hill (2007). Robert and Casella (2010) provide very easy to read coverage of the EM algorithm.

6.7   Exercises

1. Why does having a panel structure to the data violate the distributional assumptions of maximum likelihood theory?

2. What is the essential difference between a fixed- and random-effects model?

3. What is the foremost difference between an unconditional and a conditional fixed-effects model?

4. Amend the conditional fixed-effects code in Section 6.2.3 so that it can be used to estimate conditional fixed-effects logistic regression.

5. Write the needed functions to wrap the random-intercepts model above in an `ml_` style model-fitting function.

6. Use the EM algorithm for a normal regression model with missing values for one of more predictors. The use of the *norm* package from CRAN is acceptable.

7. Rewrite the EM code to use `while` instead of `for`.

7

Model Estimation Using Simulation

7.1 Simulation: Why and When?

Simulation has been part of statistics since near its very beginnings. However, due to the memory requirements needed to perform all but the most elementary simulations, the method was not used until recently. Markov chain sampling, for example, began with Metropolis et al. (1953) as an application in physics. It was later refashioned as a statistical sampling method by Hastings (1970). The sampling method he proposed became known as the Metropolis–Hastings sampling algorithm. We discuss the method and provide full working code for estimating a Bayesian Poisson model in the final section of the chapter. A further advance to the methodology was made by statisticians Stuart and Donald Geman in 1984 when they published a sampling method based on the Metropolis–Hastings algorithm (Geman and Geman, 1984). They named the method Gibbs sampling, after Josiah Gibbs of Yale University, one of the leading physicists and engineers in the 19th century. In 1863 Gibbs was awarded the first PhD in engineering in the United States. Gibbs is perhaps best known for his work in developing the area of statistical mechanics with Ludwig Boltzman and James Clerk Maxwell, as well as being the inventor of vector calculus. Today the Metropolis–Hastings sampling algorithm and Gibbs sampling are the two foremost methods of sampling used in Bayesian modelling. Here, we cover only the former.

The sampling method proposed by Hastings did not gain popularity until computers were powerful enough to run the many iterations needed for the algorithms to find the appropriate posterior distribution for a given model. Some work had been done using mainframe computers, but personal computers were not widely available until the 1980s. The first IBM PC was not released until August 1981.

In Bayesian modelling, the posterior distributions are obtained as the product of the likelihood and prior distributions. It is from the posterior distribution that model coefficients, standard deviations, and confidence/credible intervals are determined. In the early days of Bayesian modelling, the prior distribution was nearly always the conjugate of the likelihood. The conjugate distribution, as will be discussed in the final section of the chapter, has the same form as its associated probability, and therefore likelihood, function. This allows for a relatively easy calculation of the posterior. However, when the conjugate is not appropriate, or available, for a given PDF, calculations of the posterior can become very difficult, if not in practice impossible. The Metropolis–Hastings method uses a Markov Chain sampling procedure, explained later, which employs a large number of iterations before stabilizing at the posterior distribution of a model parameter; i.e., a coefficient or scale parameter. Each parameter of a model has its own posterior distribution from which the relevant summary statistics can be abstracted. In any case, this procedure requires substantial computing memory and speed.

When Bayesian modelling was limited for practical purposes to models based on conjugate distributions, many leading statisticians found fault with the entire procedure, and in particular with the subjectivity with which prior distributions were brought into the modelling process. But when computers became powerful enough to fully employ methods like Metropolis–Hastings and Gibbs sampling, such objections weakened substantially.

Bootstrap re-sampling was initiated by Bradley Efron in 1979, and became a popular method for assessing model distributional assumptions from the early 1990s (Efron, 1979). The acceptance of jackknife and bootstrap sampling, however, made it easier for statisticians to accept Bayesian methodology, which did not begin to become in use until shortly before 2000. Even with increased computing power, and an acceptance of sampling as acceptable statistical methods, finding an appropriate Bayesian posterior distribution for more complex models, and in particular hierarchical models, was still difficult, and only a few statistical packages existed for those wishing to engage in Bayesian modelling.

With the turn of the century came WinBUGS, MLwiN, and R software. Versions of these applications were available in beta and development form earlier than this, but with 2000 came packages for general use. MLwiN is a hierarchical models application with Bayesian modelling capability that is developed at the University of Bristol in the U.K., and WinBUGS is mutually developed at Imperial College School of Medicine London and by MRC Biostatistics in Cambridge, U.K. Based on Gibbs sampling, it is currently the most well-used Bayesian application. SAS also has Bayesian modelling capability, having recently added it to the Genmod Procedure, SAS's primary GLM and GEE modelling software. The procedure uses a Gamerman sampling algorithm. SAS has other Bayesian procedures as well. R developers have gradually been adding Bayesian capabilities to R, but many R programmers prefer to write their own functions, including their own Metropolis–Hastings and Gibbs algorithms. Given the complexity of a generic Bayesian function R programmers prefer to develop software for the particular modelling task at hand. In the final section of this chapter, we develop a complete Metropolis–Hastings algorithm, which readers can then expand on for their own needs.

First though we show how sampling and what has become known as *Monte Carlo* methodology develops. We begin by creating synthetic statistical models which we have used in the text for assessing the distributional properties of various models. We shall look with more detail here to how such synthetic models are developed. Following an examination of a single synthetic model, we embed synthetic models within a sampling or Monte Carlo algorithm which turns point estimate coefficients and parameters into random variables. The mean, standard deviation, and confidence intervals are calculated in the normal manner. We next turn to sampling real model data from estimated coefficients obtained using a maximum likelihood function. Finally, in the last section, we discuss and develop an annotated Metropolis–Hastings Poisson model.

7.2 Synthetic Statistical Models

Synthetic models are valuable for testing model assumptions and the bias of model statistics. We shall demonstrate in this section how synthetic models may be used to test our statement in Section 4.10 that the Pearson Chi2 dispersion statistic is the appropriate measure or test for assessing apparent count model over-dispersion, whereas the deviance dispersion is not. In order to test this contention, we shall construct synthetic Poisson and negative binomial models, including the Monte Carlo estimates of model parameters, including coefficients, standard errors and confidence intervals. The object of the exercise is to demonstrate how to construct such models, and then how to apply them to a particular problem.

7.2.1 Developing Synthetic Models

There has been some discussion in the statistical literature regarding the use of the deviance or the Pearson Chi2 based dispersion statistic for assessing the apparent extra-dispersion inherent in count models. We say "apparent" because it may be the case that the dispersion statistic indicates over-dispersion, but if the researcher who is modelling the data employs an interaction term, or converts a predictor to another scale, or performs some other operation on the model, then the apparent over-dispersion may disappear. A full discussion of this problem may be found in Hilbe (2009, 2011). We shall assume that all such checks of the data have been made and that the value of the dispersion statistic under consideration remains.

Obtaining discrete response model dispersion statistics involves dividing the respective deviance or Pearson Chi2 statistic by the model residual degrees of freedom. R's `glm` and `glm.nb` functions provide the residual deviance and residual degrees of freedom in their default output, as well as in the summary function output, but nothing regarding the Pearson Chi2 statistic is displayed. As a consequence, many researchers using R simply employ the deviance dispersion statistic to check for possible count model over-dispersion. There were other reasons why statisticians believed that the deviance disper-

sion is a good measure for assessing excess correlation in count data. For true
binomial models, e.g., grouped logistic regression models, both the Pearson
and deviance dispersion statistics produce a value of 1.0. Moreover, until recently the deviance statistic was itself used as a goodness-of-fit statistic for
Poisson models. The deviance statistic is Chi2 distributed with degrees of freedom equal to the number of observations in the model minus the number of
predictors, including intercept. In addition, an Analysis of Deviance Table has
been a popular means to assess the worth of model predictors. Therefore, the
deviance appears to have a central role in Poisson model fit analysis. However,
as it turns out, the deviance dispersion is biased for count models with respect
to correlation, displaying over-dispersion in the data when in fact there may
not be. Only the Pearson dispersion has a value of 1.0 for a well-fitted count
model.

How do we know this? We create synthetic models that are designed to
have no violations of model assumptions. If we create a *true* Poisson model,
for example, and hold that either the deviance or Pearson Chi2 dispersion is
acceptable for assessing extra-dispersion, then we should expect the dispersion
statistics to approximate 1.0. The Pearson dispersion statistic may be calculated using the saved statistics with `glm` or `glm.nb`. Assuming a model name
of *mymod*, we display the Pearson Chi2 statistic and its associated dispersion
using the code presented in Section 4.10.

A synthetic model is based on the inverse transformation of the model
probability distribution function, or PDF. In the case of the Poisson model
this inverse transformation is $\exp(x\beta)$. This value is used as the Poisson mean.

First we generate explanatory predictors for the model as pseudo-random
uniform numbers. Next, we specify coefficient values to be associated with the
predictors, including an intercept. The sum of the coefficient-predictor terms
is the linear predictor, $X\beta$ (`xb` in the code). The inverse-link transformation
is applied to the linear predictor, resulting in the mean, or μ. In the code
below we designate the mean as *exb*. The vector of mean values is then put
into the Poisson generator, together with the number of observations given the
model. This results in a vector of random Poisson variates, *py*. Modelling *py*
on the uniform predictors we specified produces a synthetic model with near
identical values to the assigned coefficient values, together with associated
model statistics. The example below is given an intercept of 2 and respective
coefficient values of 0.75 and −1.25. The synthetic model comprises 50,000
observations.

```
> # Synthetic Poisson
> # =====================
> # syn.poisson.r
> set.seed(1)
> nobs <- 50000
> x1 <- runif(nobs)
> x2 <- runif(nobs)
> xb <- 2 + .75*x1 - 1.25*x2 # linear predictor
```

```
> exb <- exp(xb)                # mean
> py <- rpois(nobs, exb)        # random Poisson variates
> poireg <-glm(py ~ x1 + x2, family = poisson)
> coef(summary(poireg))
```

```
              Estimate   Std. Error    z value Pr(>|z|)
(Intercept)  2.0025563 0.004684840    427.4546        0
x1           0.7469615 0.006261905    119.2866        0
x2          -1.2551332 0.006392715 -196.3380          0
```

The Pearson Chi2 dispersion is then

```
> sum(residuals(poireg, type="pearson")^2) / poireg$df.residual
```

```
[1] 0.9951837
```

and the deviance dispersion is

```
> with(poireg, deviance/df.residual)
```

```
[1] 1.040959
```

We notice that the deviance dispersion is 1.041, some 4.1% greater than unity. The Pearson dispersion approximates 1.0. The fact that we used data comprising 50,000 observations indicates that the estimates should be close to their true values. However, given that pseudo-random numbers are being used to create the model, and that each run produces another set of coefficient values and dispersion statistics, albeit close to the values displayed here, it is possible that random variation has resulted in the apparent high deviance-dispersion value.

Before dealing with the possibility that the synthetic model we generated above has not produced true values, we develop a synthetic binomial-logit model, or synthetic logistic regression. A synthetic grouped logit model is created in the same manner as the synthetic Poisson model was developed above. The only differences are the inverse transform, or inverse link function, which is $1/(1 + \exp(-xb))$ or $\exp(xb)/(1 + \exp(xb))$, and the creation of a binomial denominator. The identical coefficient values are specified.

It should be noted that binary response (1/0) binomial models are not extra-dispersed. They may be implicitly extra-dispersed if, when converted to grouped format, they are shown to be extra-dispersed (see Hilbe, 2009). But as binary response models they are not over- or under-dispersed. We also need to remind readers that R has a unique manner of setting up grouped binomial models. The binomial numerator is subtracted from the denominator to form a not-y variable, which is entered into the algorithm as combined dual term. In Chapter 4, we show how to develop grouped logit models that employ the traditional format. For our purposes here, though, it makes no difference; the statistical results are identical.

We need to generate a binomial denominator before inserting it into the algorithm. Usually the data has such a variable, indicating how many observations have the identical covariate pattern. In the code below we create a denominator having 10,000 observations each of the values 100, 200, ..., 500. No numerator value exceeds an associated denominator value, d. A table of denominator values is displayed below the model statistics.

```
> # Synthetic Grouped Logit
> # ==============================
> # syn.glogit.r
> nobs <- 50000
> set.seed(1)
> x1 <- runif(nobs); x2 <- runif(nobs)
> d <- rep(1:5, each=10000, times=1)*100 # binomial denominator
> xb <- 2 + .75*x1 - 1.25*x2               # linear predictor
> exb <- 1/(1+exp(-xb))                    # mean
> by <- rbinom(nobs, size = d, p = exb)  # logit variates
> dby = d - by                            # set up y and not-y
> gby <- glm(cbind(by, dby) ~ x1 + x2,
+             family = binomial)
> coef(summary(gby))
```

```
                Estimate  Std. Error    z value Pr(>|z|)
(Intercept)   2.0003226 0.001972347 1014.1837        0
x1            0.7484219 0.002520616  296.9203        0
x2           -1.2473561 0.002550580 -489.0480        0
```

The Pearson Chi2 dispersion is then

```
> sum(residuals(gby, type="pearson")^2) / gby$df.residual
```

```
[1] 0.9944635
```

and the deviance dispersion is

```
> with(gby, deviance/df.residual)
```

```
[1] 1.000861
```

Notice that both the Pearson and deviance dispersion values approximate 1.0, confirming the statement we made earlier. The code may be made more brief by combining terms. This is the case with much of R code. The problem, however, is in being able later to interpret what was combined. For example, the coefficients, predictors, and inverse transform we defined above may be combined into a single term within the rpois function. We add the confint function, which calculates model confidence intervals.

```
> confint(gby)
```

```
                  2.5 %       97.5 %
(Intercept)  1.9964576   2.0041891
x1           0.7434818   0.7533624
x2          -1.2523556  -1.2423575
```

Note that a message appears directly under the `confint` function when it is being run. The user is informed that profile confidence intervals are being developed. As noted in Section 2.4, profile intervals differ from Wald confidence intervals. Wald confidence intervals may be obtained using the `confint.default` function. In either case, however, the use of standard errors in statistical modelling is essential to the interpretation of coefficients. True synthetic models will have profile confidence intervals that will be near identical to model-based confidence intervals.

```
> confint.default(gby)   # Wald

                  2.5 %       97.5 %
(Intercept)  1.9964569   2.0041883
x1           0.7434815   0.7533622
x2          -1.2523551  -1.2423570
```

7.2.2 Monte Carlo Estimation

We now return to the query regarding how we could be sure that the statistics displayed in the synthetic Poisson model output accurately reflect the true underlying values inherent in the data. Recall that we asked if the dispersion statistics, which were not assigned to the model, but rather were generated from the model, might have values that significantly vary from what we would expect in a true equi-dispersed Poisson model.

The accuracy of calculated statistics such as those that are produced in a synthetic Poisson or logit regression may be determined using a technique referred to as Monte Carlo analysis. Monte Carlo techniques entail that a large number of synthetic models are run, with the mean values of the statistics of interest being calculated at the end. The fact that the coefficients we assign to the model are near identical to the mean values displayed in the Monte Carlo results guarantee that the mean values of the statistics of interest accurately reflect values of the statistics for a true model. Thus if the coefficients of a Monte Carlo run are the same as the values we assigned the model, we can be sure that the values of the dispersion statistics are accurate as well.

The code for a Monte Carlo Synthetic Poisson is provided below. The results are displayed directly under the code. The same synthetic Poisson model we ran earlier is executed 1,000 times, with the mean values of the coefficients and both dispersion statistics saved as vectors of values. That is, the procedure generates 1,000 values for each coefficient as well as for each dispersion statistic, saving them separately for later analysis. As before, the model itself consists of 50,000 observations. With each run, a new 50,000-observation dataset is generated.

```
> # Monte Carlo Estimation: Synthetic Poisson
> # =====================================
> # sim.poi.r
> set.seed(1)
> mysim <- function() {
+   nobs <- 50000
+   x1 <- runif(nobs)
+   x2 <- runif(nobs)
+   xb <- 2 + .75*x1 -1.25*x2
+   exb <- exp(xb)
+   py <- rpois(nobs, exb)
+   poisim <- glm(py ~ x1 + x2, family=poisson)
+   pr <- sum(residuals(poisim, type="pearson")^2)
+   prdisp <- pr/poisim$df.residual
+   dvdisp <- with(poisim, deviance/df.residual)
+   beta    <- poisim$coef
+   list(prdisp, dvdisp, beta)
+ }
```

We now perform the simulations using the following code.

```
> reps <- 1000
> B <- replicate(reps, mysim())
```

Having performed the simulations, we are now in a position to assess the distributions of the dispersion statistics. The Pearson dispersion is

```
> summary(unlist(B[1,]))
```

```
   Min. 1st Qu.  Median    Mean 3rd Qu.    Max.
 0.9715  0.9952  0.9998  0.9998  1.0050  1.0190
```

and the deviance dispersion is

```
> summary(unlist(B[2,]))
```

```
   Min. 1st Qu.  Median    Mean 3rd Qu.    Max.
  1.015   1.041   1.045   1.045   1.050   1.064
```

We can obtain histograms of the parameter estimates using the following code
(Figure 7.1).

```
> beta.hat <- do.call(rbind, B[3,])
> par(mfrow=c(1,3), mar=c(5,4,1,2), las=1)
> for (i in 0:2)
+   hist(beta.hat[,i+1], breaks = 50, main = "",
+        xlab = bquote(paste("Simulated ", beta[.(i)])))
```

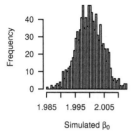 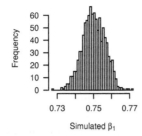 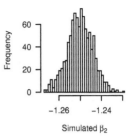

FIGURE 7.1
Histograms of simulated parameter estimates. The x-axis labels are unnecessarily ornate.

The results closely resemble the values given in the single run model. It is clear that a "true" Poisson model for these data – one for which the mean and variance are identical – has a Pearson Chi2 dispersion statistic of 1.0, and a deviance dispersion of 1.0455, some 4.5% higher than unity. Our interpretation is that Poisson models are apparently over-dispersed, based on the deviance dispersion. It may be the case, in general, that converting a term in the model to another scale, or perhaps adding an interaction term, may result in a model with a Pearson dispersion value of 1.0. However, if no such operations are able to correct the model, then the data are truly Poisson over-dispersed.

The usual manner of handling over-dispersed Poisson data is by using a negative binomial (NB2) model. Other types of models may be more appropriate for a given data situation, but researchers usually apply a negative binomial model to the data as a first attempt at accommodating model over-dispersion. If a model is under-dispersed, i.e., the Pearson dispersion statistic is less than 1, then a negative binomial model is not appropriate. Researchers typically use a generalized Poisson, a generalized negative binomial, or a hurdle model for such data.

We have elaborated a bit on these models for the purpose of demonstrating how synthetic models may be useful in determining the statistical consequences of the distributional assumptions of a given model. In the above case, we used Monte Carlo techniques to determine the appropriate statistic to use for assessing Poisson over-dispersion. We may use the same techniques for more complex models as well, e.g., a negative binomial model.

There are a number of different types of negative binomial models. We have been using the traditional parameterization, which is used to model otherwise over-dispersed Poisson models. This parameterization is often referred to as an NB2 negative binomial (see Hilbe, 2011); we shall simply refer to it as a negative binomial model.

The negative binomial is a mixture of Poisson and gamma distributions. The gamma scale parameter in the mixture becomes the negative binomial scale parameter, or heterogeneity parameter, which is used to adjust for Poisson over-dispersion. Such over-dispersion indicates more correlation in the data than is allowed by the distributional assumptions of the model. Negative binomial models may be over-dispersed, or under-dispersed.

The negative binomial scale parameter is traditionally parameterized such that a value of 0 is a Poisson model. Increasing values indicating more dispersion, or correlation, in the data. This parameterization of the negative binomial scale parameter is usually referred to as *alpha*. Since a Poisson model has no extra correlation, i.e., the mean and variance are identical, *alpha* = 0. Values of *alpha* greater than 2 usually indicate that there was substantial over-dispersion or correlation in the Poisson data.

R's `glm` and `glm.nb` functions, the latter being part of the *MASS* package (Venables and Ripley, 2010) that comes with the default download of the software, are unique in that they parameterize the negative binomial scale as having an inverse relationship with the amount of correlation in the data. The parameter is called *theta* (θ), and a negative binomial with $\theta = 0$ is infinitely over-dispersed, and a model where $\theta =$ inf indicates a Poisson model with no extra-dispersion. Care must be taken when comparing negative binomial model results using other major software, which use *alpha*, and when using R's `glm` and `glm.nb` functions. In Chapter 5 we demonstrated developing a two-parameter maximum likelihood model with a negative binomial having *alpha* as the scale parameter.

We provide below a Monte Carlo synthetic negative binomial model using the same number and value for the coefficients. The point is to check the dispersion statistics to see if a "true" negative binomial has the same values. We would expect they do. However, even though the Pearson Chi2 dispersion is expected to be 1.0 regardless of the value of *alpha* given the model, we expect the deviance dispersion to vary depending on the value of alpha. Lower values have a lower deviance dispersion; higher values of *alpha* have a greater deviance value — up to a limit. For these data and coefficient values, the deviance dispersion ranges from a low of approximately 0.7 ($\alpha = 0.1$) to a high of 1.14 ($\alpha = 0.9$), with values more extreme making no important difference.

We use 1000 iterations. We specify `alpha = 0.5`, which is the same as $\theta = 2$. Note that alpha is inverted in the code so that R's `glm.nb` function produces the scale value that we specify.

```
> # Monte Carlo Estimation: Synthetic Negative Binomial
> # =====================================================
> # sim.nb2.r
> library(MASS)
> mysim <- function() {
+    nobs <- 50000
+    x1 <-runif(nobs)
+    x2 <-runif(nobs)
+    xb <- 2 + .75*x1 - 1.25*x2
```

```
+    a <- .5                      # alpha
+    ia <- 1/.5                   # theta
+    exb <- exp(xb)               # log link
+    xg <- rgamma(nobs, a, a, ia) # gamma variates
+    xbg <-exb*xg                 # means
+    nby <- rpois(nobs, xbg)      # Poisson variates
+    nbsim <-glm.nb(nby ~ x1 + x2) # model
+    alpha <- nbsim$theta
+    pr <- sum(residuals(nbsim, type="pearson")^2)
+    prdisp <- pr/nbsim$df.residual
+    dvdisp <- nbsim$deviance/nbsim$df.residual
+    beta <- nbsim$coef
+    list(alpha, prdisp, dvdisp, beta)
+ }
```

As before, we run the simulations by

```
> set.seed(1)
> reps <- 1000
> B <- replicate(reps, mysim())
```

Having performed the simulations, we are now in a position to assess the distributions of the dispersion statistics. The estimates of *alpha* are

```
> summary(unlist(B[1,]))
```

Min.	1st Qu.	Median	Mean	3rd Qu.	Max.
0.4875	0.4973	0.4997	0.4999	0.5024	0.5122

The Pearson dispersion is

```
> summary(unlist(B[2,]))
```

Min.	1st Qu.	Median	Mean	3rd Qu.	Max.
0.9706	0.9936	1.0000	0.9999	1.0060	1.0310

and the deviance dispersion is

```
> summary(unlist(B[3,]))
```

Min.	1st Qu.	Median	Mean	3rd Qu.	Max.
1.095	1.098	1.099	1.099	1.099	1.102

We can again obtain histograms of the parameter estimates using the same code as before (Figure 7.2), noting that the parameter estimates are retained in B[,4] instead of B[,3].

More complex models may be developed of course, including two-part mixture models such as zero-inflated count models, ordered and unordered slopes models, finite mixture models, fixed, random and mixed effects models, and so forth. A relatively simple two-part synthetic zero-inflated Poisson algorithm is displayed below. The *pscl* package must be installed for this code to work.

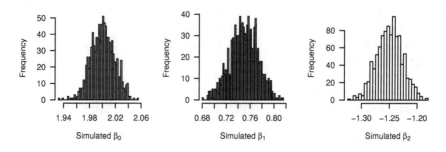

FIGURE 7.2
Histograms of simulated parameter estimates for the negative binomial regression simulation exercise.

```
> # Synthetic Zero-inflated Poisson with logit component for 0's
> # ============================================================
> # syn.zip.r
> library(MASS); library(pscl)
> set.seed(1)
> nobs <- 50000
> x1 <- runif(nobs); x2 <- runif(nobs)
> xb <- 2 + .75*x1 - 1.25*x2              # Poisson lin predictor
> exb <- exp(xb)                          # Poisson fitted values
> poy <- rpois(nobs, exb)                 # Poisson
> pdata <- data.frame(poy, x1, x2)        # Create dataframe
> pi <- 1/(1+exp(-(.9*x1 + .1*x2 + .2)))  # Logit probabilities
> pdata$bern <- runif(nobs) > pi          # Filter
> zy <- pdata$bern * poy                  # Mixture
> zip <- zeroinfl(zy ~ x1 + x2 | x1 + x2,
+                    dist = "poisson",
+                    data = pdata)
> coef(summary(zip))
```

Confidence intervals for both components of the model may be obtained using the **confint** function. Note that profile likelihood confidence intervals are not displayed since they are not an option with **zeroinfl**.

```
> confint(zip)
```

	2.5 %	97.5 %
count_(Intercept)	1.98226951	2.0129524
count_x1	0.72672558	0.7701504
count_x2	-1.27393715	-1.2284106
zero_(Intercept)	0.15508471	0.2523075
zero_x1	0.83097289	0.9631452
zero_x2	0.01384794	0.1448562

To reiterate, constructing synthetic models can be an important adjunct to understanding the importance of assumptions upon which statistical models are based. They may also be used to test the application of statistics that can be generated on the basis of a specified model, or set of models. We demonstrated this when comparatively evaluating the deviance and Pearson Chi2 dispersion statistics for count models, as well as comparing them for use with binomial grouped logistic regression models. A host of other tests may be performed as well, which lead to a better understanding of the statistical models we employ in research. This type of testing appears to have first been used by Hilbe and Linde-Zwirble (1995), although the authors would welcome earlier references. Refer to Gelman and Hill (2007) and Hilbe and Linde-Zwirble (1995); Hilbe (2010, 2011) for a further discussion.

We demonstrate the use of a simulation algorithm, sim, found in Andrew Gelman's *arm* package, located on CRAN. Following estimation of medpar data using a Poisson regression, the two binary coefficients, hmo and white are treated as random variables. Empirical distributions are created for them. We take the mean, the standard deviation and apply the quantile function to obtain simulated coefficients, standard errors, and confidence intervals of each predictor.

```
> library(msme)
> data(medpar)

> library(arm)
> # fit initial model to get coefficients
> fit.1 <- glm(los ~ hmo + white,
+              family = poisson,
+              data = medpar)
> fit.1$coef

(Intercept)        hmo        white
  2.4822518  -0.1415782  -0.1908900

> # simulation 1000 "random" values of each coefficient
> n.sims <- 1000
> sim.1 <- sim(fit.1, n.sims)
> pcoef <- coef(sim.1)

> # get stats for hmo
> hmo.coef <- pcoef[,2]
> summary(hmo.coef)

    Min.   1st Qu.   Median    Mean   3rd Qu.    Max.
-0.22820  -0.15810  -0.14160  -0.14170  -0.12630  -0.06424

> quantile(hmo.coef, p = c(0.025, 0.975))
```

```
          2.5%         97.5%
-0.18818771 -0.09115243

> # get stats for white
> white.coef <- pcoef[,3]
> summary(white.coef)

   Min. 1st Qu.  Median    Mean 3rd Qu.    Max.
-0.2693 -0.2111 -0.1904 -0.1911 -0.1717 -0.1101

> quantile(white.coef, p = c(0.025, 0.975))

       2.5%        97.5%
-0.2438868 -0.1345252
```

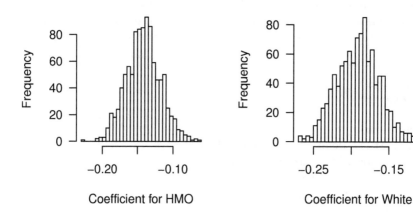

FIGURE 7.3
Histograms of simulated parameter estimates for the simulated Poisson regression simulation exercise using *arm*.

7.2.3   Reference Distributions

We conclude this section with some code that demonstrates that choosing
a reference distribution for testing a random-effects model as opposed to a
pooled model is a tricky problem. First, we obtain the data for the model.

The `rwm5yr` dataset is based on the German Health Registry from the
years 1984–1988. It is a panel dataset, but not every patient was observed for
each of the 5 years. The response variable is `docvis`, the number of days a
patient visits a physician during the calendar year. The predictors are `age`, a
continuous variable from 25–64; `outwork`, a binary variable with 1 = patient

out of work, 0 = employed during year; `female`, a binary variable with 1 = female, 0 = male; `married`, a binary variable with 1 = married, 0 = single; and `id`, the patient ID number.

```
> library(COUNT)
> library(nlme)
> data(rwm5yr)
> rwm5yr <- rwm5yr[,c(1,2,6:9)]
> for(i in 1:6) class(rwm5yr[,i]) <- "numeric"
> rwm5yr$id <- factor(rwm5yr$id)
```

We can now fit the random intercept model and the pooled model and, in theory, compare them using the likelihood ratio test, for which the reference distribution is the Chi2 distribution with 1 degree of freedom.

```
> test.lme <- lme(docvis ~ age + outwork + female + married,
+                    random = ~1 | id,
+                    method = "ML",
+                    data = rwm5yr)
> test.lm <- lm(docvis ~ age + outwork + female + married,
+                 data = rwm5yr)
> anova(test.lme, test.lm)
```

```
         Model df    AIC       BIC    logLik    Test  L.Ratio
test.lme     1  7 121893.2 121948.4 -60939.59
test.lm      2  6 124311.1 124358.4 -62149.53 1 vs 2 2419.89
         p-value
test.lme
test.lm  <.0001
```

However, we can also develop a reference distribution for this particular case in the following way. First, we simulate from the null model — here, the pooled model — a given number of times. We then fit both models to the simulated data, for which we know that the null model is correct. Hence, any appearance of improvement of the random effects model is purely due to chance.

```
> reps <- 1000
> new.y <- simulate(test.lm, nsim = reps, seed = 100)
> new.rwm <- rwm5yr
> should.be.chi2 <-
+    sapply(new.y,
+           function(x){
+               out <- try({
+               new.rwm$docvis <- x
+               new.lm <- update(test.lm, data = new.rwm)
+               new.lme <- update(test.lme, data = new.rwm)
+               anova(new.lme, new.lm)$L.Ratio[2]})
```

```
+              if (class(out) == "numeric") return(out)
+              else return (NA)
+        })
```

We can now use the reference distribution in two ways: first, we can compare it with the observed critical value, which in this instance is much greater than any of these simulated values. Note from Figure 7.4 that the maximum simulated value is 15, and the observed value from the output on the previous page is greater than 2000. Second, we can compare the distribution of the simulated values with the theoretical reference distribution, as per the following code (see Figure 7.4). The reference distribution is a very poor fit, but here the poor fit does not matter at all.

```
> par(mfrow=c(1,2), mar=c(4,4,2,1), las=1)
> qqplot(should.be.chi2, rchisq(10000, df = 1),
+      xlab = "Simulated Dist.", ylab = "Theoretical Dist.")
> abline(0, 1)
> plot(ecdf(pchisq(should.be.chi2, df=1)),
+      main = "Empirical Reference CDF")
```

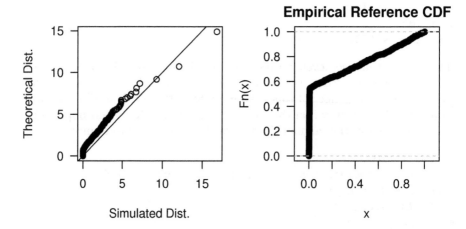

FIGURE 7.4
Comparison of simulated and reference distribution for the null hypothesis that the pooled model is just as good as the random effects model for the rwm5yr data.

7.3 Bayesian Parameter Estimation

As we saw in Chapter 6, there are a number of models which preclude the use of standard maximum likelihood methods for the estimation of model parameters. When data are structured in panels, observations cannot generally be regarded as independent. There is likely greater correlation within panels than between panels. In fact, when maximum likelihood, or better quasi- or pseudo-maximum likelihood, methods are used for the estimation of panel models, panels of observations are assumed to be independent, whereas the observations within panels are correlated. The model is adjusted by various means to accommodate for the correlation within panels. Generalized Estimating Equations (GEE), for example, employ various types of correlation structures into the distributional variance of the model in order to adjust for the within-panel correlation. The result is a model that estimates unbiased parameters. Random-, fixed-, and mixed-effects models also aim to provide appropriate adjustment for panel data.

We also have previously observed that even with adjustments to a maximum likelihood algorithm, at times such methods are still not appropriate for the unbiased estimation of model parameters. In particular, application of an EM algorithm or the use of quadrature now finds widespread use in commercial statistical software for the estimation of hierarchical models. However, even these methods may still not be satisfactory for models based on complex distributions. And even where quadrature, for example, may be used, it may be accompanied with considerable convergence difficulties as well as an unsatisfactory handling of variability.

In this chapter we have demonstrated how sampling can be used to better understand models based on maximum likelihood estimation. We began by developing a synthetic Poisson variable that was structured such that it had two inherent predictors with specifically defined values. We used the `runif` function to define two random uniform variates and the `rpois` function to generate a random Poisson number having a mean defined as the exponentiation of `xb`, a linear predictor formed from the random uniform numbers.

```
> nobs <- 100
> x1 <- runif(nobs)
> x2 <- runif(nobs)
> xb <- 2 + .75*x1 -1.25*x2
> exb <- exp(xb)
```

This procedure allows us to create a synthetic Poisson model with user-defined coefficients. For a negative binomial model, the same procedure may be extended to allow the data to incorporate a defined scale parameter.

Generating synthetic models like this allows researchers to create true models. The data is such that all of the distributional assumptions of the model are met. Real data, and statistics associated with the real-data models, can then be compared to true versions. By developing a synthetic true Poisson model we can determine that the Pearson-based dispersion statistic is the appropriate statistic to use for assessing Poisson extra-dispersion. It is also the appropriate statistic to use for assessing negative binomial extra-dispersion. On the other hand, either the Pearson or deviance dispersion may be used to determine grouped or proportional binomial-logit extra-dispersion.

The algorithm used to create a synthetic Poisson model, for example, can be embedded into a covering algorithm that repeatedly creates estimates of the coefficients and specified ancillary statistics, saving the results of each estimate iteration. In this manner we in fact make a major change to how we understand obtaining parameter estimates and values for associated statistics.

Recall that the coefficients of a synthetic model are not exactly the values as specified at the start of the algorithm. They come back fuzzy, with a variability that can be measured by the standard errors of the coefficients, for example. Looked at in this manner, the vectors of saved coefficients and associated model parameters are each random variables. The mean of the vector of coefficients and other parameters and related statistics are their estimated Monte Carlo values.

Earlier in this chapter we created Monte Carlo estimates for both synthetic Poisson and negative binomial models. This procedure allows us to more accurately assess the distributional assumptions of the respective models, and to compare the known statistics with real model data.

We started the Monte Carlo algorithm for synthetic Poisson data by specifying point estimates for the coefficients. By re-estimating the synthetic coefficient and dispersion values and saving them as vectors of values, we took their mean, and also their standard deviation and quantiles, or credible intervals, which are similar to confidence intervals, but without the associated asymptotics. We also plotted a histogram for the intercept and each predictor, x1 and x2, so we could more easily observe the distribution of the coefficients. We may ask for such a model, is the assumption warranted that Poisson regression coefficients are normally distributed? Or, are Gaussian regression coefficients actually t-distributed? We also provided a more complex mixture model — a zero inflated Poisson which is a mixture of Poisson and binomial-logit distributions.

We next changed the focus of modelling from synthetic to real data. The object was to demonstrate that it is possible to use a simulation function to create vectors of coefficient and associated parameter values of real data. We therefore now will think of regression parameters as random variables and not as fixed values to be discovered. This is the foundation upon which Bayesian modelling is based.

There are a number of data situations which cannot be modeled using standard statistical techniques. We mentioned this earlier. Simulation, however, has been gaining in popularity as a method of estimating parameters for otherwise intractable models. Limdep, a popular commercial econometric software package, uses simulation for many of its more complex models. Bayesian models with flat or uniform priors produce the same parameter estimates as maximum likelihood, within limits. We shall demonstrate this fact in the current section. The methods are substantially different, but the results are closely the same. Lynch (2007), Gill (2009), Hilbe (2011), and Zuur et al. (2012) are all excellent resources on using R for developing Bayesian GLM regression models.

The Bayesian approach is similar to the Monte Carlo approach used in our code earlier this chapter, but extends its scope by not being based on a previously estimated model. Bayesian models use the raw data itself to estimate parameters as random variables. The distribution that is used for a parameter — and from which the coefficients, standard errors and credible intervals are estimated — is termed the posterior distribution.

The true value in Bayesian modelling is the ability to bring informative prior information into the model. The goal is to estimate a posterior distribution from the product of the likelihood and prior distributions, which many times lead to a complex distribution that must be developed using one of a variety of Markov Chain Monte Carlo (MCMC) sampling algorithms. A number of algorithms exist, but for the most part they are based on a general MCMC procedure.

We use a basic Metropolis–Hastings algorithm to demonstrate the use of a Bayesian Poisson model using real data. We should clarify the difference and use of R's **dpois** and **rpois** functions. Both are used in the Metropolis–Hastings algorithm given below. The R function **dpois** is the Poisson probability function, defined as $\mu^y \times \exp(-\mu)/y!$. If $\mu = 0.5$ and $y = 4$, then

```
> m <- 0.5; y <- 4
> m^y * exp(-m)/factorial(y)
```

```
[1] 0.001579507
```

```
> dpois(y,m)
```

```
[1] 0.001579507
```

while the Poisson log-likelihood, $y \ln(\mu) - \mu - \ln(y!)$ can be expressed as

```
> y*log(m) - m - lfactorial(y)
```

```
[1] -6.450643
```

```
> log(dpois(y, m))
```

```
[1] -6.450643
```

or, as is preferred,

```
> dpois(y, m, log=TRUE)
```

[1] -6.450643

The R function `rpois` is used for generating a given number of Poisson random numbers with a specified mean value. We have used the `rpois` function earlier in the book, but both `rpois` and `dpois` will be particularly important in the code given below.

We shall use the same `medpar` data that has been used for several other examples in the text to display the logic of Bayesian modelling using the Metropolis–Hastings algorithm. There are many variations of the algorithm, but the fundamental logic is the same for them all. Of course, other sampling algorithms exist and are used with equal and sometimes superior efficacy. However, this algorithm is generally regarded as the traditional or standard MCMC algorithm, so is used for our example.

```
> library(msme)
> data(medpar)
```

The response variable is `los`, or hospital Length Of Stay, a discrete count variable ranging in values from 1 to 116 days. To keep the example as simple as possible, we shall use only one binary (1/0) predictor, `hmo`, which indicates whether the patient belongs to a Health Maintenance Organization (1) or is a private pay patient (0). The appropriate model for a count variable is Poisson, unless there is extra-dispersion in the data. We will not concern ourselves with dispersion for the example.

We first model the data using the `glm` function, which employs a version of maximum likelihood to calculate parameter estimates. We do this simply to compare it with the results we obtain from the Bayesian model. The log-likelihood function is also calculated.

```
> MLpoi <- glm(los ~ hmo, family = poisson, data = medpar)
> summary(MLpoi)

Call:
glm(formula = los ~ hmo, family = poisson, data = medpar)

Deviance Residuals:
    Min       1Q   Median       3Q      Max
-3.6792  -1.7763  -0.6795   0.8803  18.8406

Coefficients:
             Estimate Std. Error z value Pr(>|z|)
(Intercept)  2.310436   0.008888 259.952  < 2e-16 ***
hmo         -0.150148   0.023694  -6.337 2.34e-10 ***
---
Signif. codes:  0 '***' 0.001 '**' 0.01 '*' 0.05 '.' 0.1 ' ' 1
```

(Dispersion parameter for poisson family taken to be 1)

```
    Null deviance: 8901.1  on 1494  degrees of freedom
Residual deviance: 8859.5  on 1493  degrees of freedom
AIC: 14579
```

Number of Fisher Scoring iterations: 5

```
> logLik(MLpoi)
```

'log Lik.' -7287.331 (df=2)

```
> confint.default(MLpoi)
```

	2.5 %	97.5 %
(Intercept)	2.2930164	2.3278564
hmo	-0.1965864	-0.1037091

The first function needed in the code specifies the log-likelihood of the model. Since the response variable, los, is a count, we use a Poisson log-likelihood. Recall that the coefficients we are attempting to determine are each random parameters. If this were a normal or Gaussian model we would have to have a parameter for the variance, sigma2. The negative binomial would likewise have a parameter for the scale. This code is not packaged in a function, but with a little work can be made to be so. We re-use the following six functions from Chapter 5.

```
> jll.poisson <- function(y, mu, m, a) {
+     dpois(x = y, lambda = mu, log = TRUE)
+ }
> jll <- function(y, y.hat, ...) UseMethod("jll")
> predict.expFamily <- function(object, b.hat, X, offset = 0) {
+     lin.pred <- as.matrix(X) %*% b.hat + offset
+     y.hat <- unlink(object, lin.pred)
+     return(y.hat)
+ }
> Sjll <- function(b.hat, X, y, offset = 0, ...) {
+     y.hat <- predict(y, b.hat, X, offset)
+     sum(jll(y, y.hat, ...))
+ }
> unlink <- function(y, eta, ...) UseMethod("unlink")
> unlink.log <- function(y, eta, m=1, a=1) exp(eta)
```

We now set up the objects that are specific to this particular problem, including the response variable, model matrix, and class information.

```
> mh.formula <- los ~ hmo
> family <- "poisson"
> link <- "log"
> mf <- model.frame(mh.formula, medpar)
> y <- model.response(mf, "numeric")
> X <- model.matrix(mh.formula, data = medpar)
> class(y) <- c(family, link, "expFamily")
```

Hopefully we should now be able to evaluate the summed joint log-likelihood at a specific value of the parameter estimates.

```
> Sjll(c(1,0), X, y)
```

```
[1] -15613.25
```

Next we develop a function to provide information about the prior distribution. The `BLogPrior` function specifies the log of the prior probability distributions for the intercept and slope. A very weak or diffuse uniform prior is given to both parameters, which are called `f.prior.a` and `f.prior.b` respectively. Recall that a uniform distribution has two parameters, a and b, within the extremes of 0 and 1. If we believe that every value within the range of a and b has an equal or uniform value, use of the uniform distribution makes sense. Parameters outside the range of a and b have a probability of 0. Within the range, parameters have a value of $1/(b-a)$. A wide range leads to its designation as diffuse or weak.

```
> BLogPrior <- function(theta){
+    alpha <- theta[1]
+    beta <- theta[2]
+    fprior.a <- dunif(alpha, -25, 30)
+    fprior.b <- dunif(beta, -25, 30)
+    fprior <- fprior.a * fprior.b # a, b independent
+    if (fprior > 0) return(log(fprior))
+ }
```

We could write more efficient code, e.g., `fprior <- dunif(theta, -25, 30)`; however, the approach that we used seems a better illustration of the principle.

We then specify that the MCMC algorithm will iterate 50,000 times in searching for the appropriate posterior distribution. We in fact do not need that many iterations. When informative priors are used in a model, 50,000 iterations will rarely be sufficient.

Matrices are then defined that will hold the columns of coefficients. The important line defines `Theta.t`, which has a dimension defined as the number of iterations plus 1, by 2, or dim(n,2). The extra 1 is for `current.Theta`, defined as c(0,0), which is the first iteration, or `Theta.t[1,]`. The second dimension is for parameters *alpha* and *beta*.

```
> nT <- 50000
> Theta.t <- matrix(nrow = nT+1, ncol = 2)
> Theta.star <- vector(length = 2)
> current.Theta <- c(1, 0)
> Theta.t[1,] <- current.Theta
> acc <- 1
```

Then we go straight to the MCMC algorithm. The sampling algorithm begins by defining a proposal distribution taken from the normal distribution. The proposal distribution is the distribution of the next draw or proposed value in the search for a parameter value. Random values are therefore taken from the normal distribution with mean based on the current value of the parameter, θ^i. The variance of the proposal distribution affects the volatility of the search. Those values are the proposed new value of the parameter, θ^p. If the proposed value passes the selection criterion given below, then it becomes the new current value, θ^{i+1}.

The value of θ^i is compared with the new proposed value θ^p in the following way. We compute

$$logR = \mathcal{P}(\theta^p|d) - \mathcal{P}(\theta^i|d) \qquad (7.1)$$

where $\mathcal{P}$ refers to the log of the posterior probability of θ, given the data. The log of the posterior probability is computed as the sum of the log-likelihood and the log of the prior distribution.

If $logR > 0$ then we accept the draw, that is, $\theta^{i+1} \leftarrow \theta^p$. If $logR$ is less than 0, then we accept the proposed value with probability $\exp(logR)$. In other words, we draw a random number uniformly distributed between 0 and 1 and compare it with $\exp(logR)$. If the random number is less than $\exp(logR)$, then we accept the proposed values; if greater, then we retain the previous values. Note that the single line of code below tests both of these conditions at the same time because $\log(u) < 0$ by definition.

Either way, we then randomly draw another proposal from a normal distribution starting from θ^i (if the proposed value was rejected) or θ^{i+1} (if the proposed value was accepted).

The code below uses the term $logU$ to compare with $logR$ for this final decision in the selection process. The iterations continue in that fashion through the entire 50,000 iterations, by which time the parameters for the intercept and slope have stabilized.

```
> # Metropolis--Hastings MCMC algorithm
> for (i in 1:nT) {
+ # Normal samples or draw for a proposal distribution
+ # We keep the proposal distributions distinct for clarity.
+    Theta.star[1] <- rnorm(1, Theta.t[acc,1], 0.1)
+    Theta.star[2] <- rnorm(1, Theta.t[acc,2], 0.1)
+ # Calculate log(R)
```

```
+      logR <-
+         (Sjll(Theta.star, X, y) + BLogPrior(Theta.star)) -
+            (Sjll(Theta.t[acc,], X, y) + BLogPrior(Theta.t[acc,]))
+ # Draw new uniform random variate
+      u <- runif(1)
+ # Compare u and r
+      if (log(u) < logR) {
+         acc <- acc + 1
+         Theta.t[acc,] <- Theta.star
+      }
+ }
>
```

It is of interest to note the proportion of values accepted. This is computed by

```
> (acc - 1) / nT
```

```
[1] 0.03612
```

Plots displaying the range of values throughout the iterations are provided in Figure 7.5. This figure shows the *burn-in* phenomenon, in which the first portion of the simulated values are discarded in the hope of minimizing the impact of the nominated start point on the outcome. In this example we use a burn-in period of 500 simulations, which is adequate for our purposes but may be too low for operational work.

```
> burn.in.length <- 500
> burn <- 1:burn.in.length
> use <- (burn.in.length+1):acc
> par(mfrow = c(2,2), mar = c(5,4,1,2), las = 1)
> plot(Theta.t[burn, 1], xlab = "Index", ylab = "alpha",
+       type= "l")
> abline(h = coef(MLpoi)[1], lwd = 2)
> plot(Theta.t[burn, 2], xlab = "Index", ylab = "beta",
+       type = "l")
> abline(h = coef(MLpoi)[2], lwd = 2)
> plot(Theta.t[use, 1], xlab = "Index", ylab = "alpha",
+       type = "l")
> abline(h = coef(MLpoi)[1], lwd = 2)
> plot(Theta.t[use, 2], xlab = "Index", ylab = "beta",
+       type = "l")
> abline(h = coef(MLpoi)[2], lwd = 2)
```

We also develop histograms of the parameters for the intercept and slope (Figure 7.6).

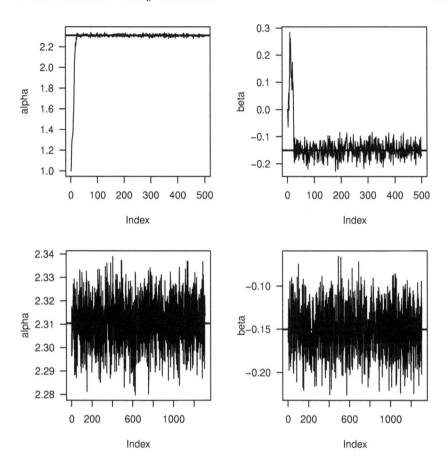

FIGURE 7.5
Parameter estimate trajectories. The left column is for the intercept, and the right column for the slope. The upper row reports the first 500 values, the burn in, and the lower row reports the balance, to be used for estimation.

```
> par(mfrow=c(1,2), mar = c(5,4,1,2), las = 1)
> hist(Theta.t[use, 1], breaks=50, main = "", xlab = "Alpha")
> hist(Theta.t[use, 2], breaks=50, main = "", xlab = "Beta")
```

Finally, the means, standard deviation, and 2.5% and 97.5% quantiles for each parameter, ignoring the first 500 simulations, are calculated and displayed.

```
> apply(Theta.t[use, ], 2, mean, na.rm = TRUE)
```

```
[1]   2.3105982 -0.1499656
```

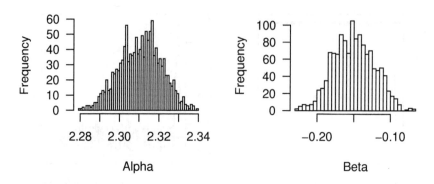

FIGURE 7.6
Histograms of estimated posterior densities of parameter estimates.

```
> apply(Theta.t[use, ], 2, sd, na.rm = TRUE)

[1] 0.01066127 0.02771301

> quantile (Theta.t[use, 1], na.rm = TRUE,
+           probs = c(.025, .975))

    2.5%      97.5%
2.289849 2.330510

> quantile (Theta.t[use, 2], na.rm = TRUE,
+           probs = c(.025, .975))

      2.5%          97.5%
-0.20226722 -0.09798302
```

We prepared a summary table comparing the maximum likelihood results from **glm** with the Bayesian results.

```
SUMMARY
Metropolis--Hastings                      Credible Intervals
                    coef        sd         2.5%        97.5%
Intercept       2.309970  0.010721    2.2899090     2.330998
hmo            -0.151302  0.028923  -0.20760979  -0.09475081


Maximum Likelihood                        Confidence Intervals
                    coef        se         2.5%        97.5%
(Intercept)     2.310436  0.008888    2.2930164    2.3278564
hmo            -0.150148  0.023694   -0.1965864   -0.1037091
```

The results are remarkably close, considering the very different manner in which the parameters and statistics are obtained. Of course, we used a simple example based on a known and simple probability and likelihood function. Priors may be represented by a variety of distributions, which, when multiplied by the likelihood, can result in very difficult posterior distributions to develop. Conjugate priors may be used for situations that are not complex, resulting in much easier posteriors to calculate.

We now demonstrate fitting the same model using the *MCMCpack* package. We select default priors for the parameter estimates, and a burn in length of 5000. Readers should consult the package documentation.

```
> library(MCMCpack)
> p.fit <- MCMCpoisson(los ~ hmo,
+                      burnin = 5000, mcmc = 100000,
+                      data = medpar)
> summary(p.fit)

Iterations = 5001:105000
Thinning interval = 1
Number of chains = 1
Sample size per chain = 1e+05

1. Empirical mean and standard deviation for each variable,
   plus standard error of the mean:

             Mean      SD  Naive SE Time-series SE
(Intercept) 2.3103 0.00871 2.754e-05      8.131e-05
hmo        -0.1501 0.02352 7.438e-05      2.223e-04

2. Quantiles for each variable:

             2.5%     25%     50%      75%  97.5%
(Intercept) 2.2931  2.3044  2.3104  2.3162  2.327
hmo        -0.1961 -0.1658 -0.1501 -0.1343 -0.104
```

Here we have discussed only the basics of the Metropolis–Hastings algorithm, For those who wish to have more details on the use of this method we suggest Zuur et al. (2012). Our presentation of the algorithm and code is adapted in part from that source. Zuur et al. discuss a normal model with both informative and non-informative priors.

7.3.1 Gibbs Sampling

Gibbs sampling is the foremost alternative to the Metropolis–Hastings algorithm we have thus far discussed. In fact, Gibbs sampling is a special version of the Metropolis–Hastings algorithm, in a similar manner as the IRLS algorithm is a version of maximum likelihood estimation. Gibbs is particularly useful when modelling multivariate models.

As we have discussed, the Metropolis–Hastings algorithm takes samples from the entire joint distribution, which can take a considerable amount of time to cover the entire parameter space. The Gibbs sampling method differs from Metropolis–Hastings in that the unknown parameters of a distribution or model are partitioned and estimated in sequence, as marginal or conditional distributions. Each parameter, or group of parameters, in this process is estimated on the basis of the other parameter values.

Suppose that we wish to estimate the parameters in the density

$$f(\theta|x) = f(\theta_1, \theta_2, \ldots, \theta_k|x) \tag{7.2}$$

Also, assume that we know the k conditional densities, that is, we know $f(\theta_1|\theta_2, \theta_3, \ldots, \theta_k|x)$, $f(\theta_2|\theta_1, \theta_3, \ldots, \theta_k|x)$, etc. After initializing each parameter with values appropriate to it, the algorithm samples from each conditional distribution in turn. That is, we start with $(\theta_1^0, \theta_2^0, \ldots, \theta_k^0)$. We then take a random draw θ_1^1 from the known density

$$\theta_1^1 \sim f(\theta_1|\theta_2^0, \theta_3^0, \ldots, \theta_k^0|x) \tag{7.3}$$

This sequence repeats until convergence. The sampling is a Markov process where the distribution at each step is independent of previous steps except for the immediately previous step.

7.4   Discussion

Calculation of a distribution by simulation is not a panacea. A posterior distribution may require that the algorithm take 100,0000 or even a million iterations before stabilization. Autocorrelation may be discovered in the iterations, which need to be thinned or adjusted. Taking every fifth or sixth iteration and discarding the rest may result in an appropriate sampling distribution with which to serve as the posterior distribution of the model. Sometimes it is better to use the mode or a trimmed mean for calculating the Bayesian coefficient and related statistics. Many considerations need to be made when the posterior is difficult to calculate. However, the algorithms found in WinBUGS, SAS and other recognized applications usually can find the proper posterior for a given data situation.

A particular complication arises when it is not possible to provide a likelihood for a model one is attempting to develop. Recently this has been addressed with the development of ABC algorithms, which is an acronym for Approximate Bayesian Computation. The method has been used with considerable success in the fields of genetics and epidemiology, although we believe it can also have use in astrostatistics and ecology, which are areas of our special interest. Several variations of the ABC algorithm have already been developed, e.g., Probabilistic ABC. It is likely that this field will develop in future years. The R package *abc* on CRAN provides an algorithm with a rather extensive modelling capability for this set of models.

7.5 Exercises

1. Our simulations to assess the accuracy of the deviance and the Pearson dispersion used quite large samples. How would our conclusions vary with much smaller samples?

2. Construct a synthetic probit model with user specified coefficients on the intercept of 0.5, and x_1 of 1.75, and x_2 of -0.8. Use random uniform variates to create the predictors. The data should have 10,000 observations.

3. Construct a 10,000 observation Monte Carlo gamma model with an inverse link. The number of predictors and their coefficients is the choice of the reader. Obtain coefficient values and a value for the Pearson dispersion statistic. Extra: Compare the Pearson dispersion statistic with the scale parameter obtained when adding the two-parameter gamma model to `ml_glm2` in Exercise 6 for Chapter 5.

4. Construct a synthetic zero-inflated negative binomial model similar to the synthetic zero-inflated Poisson model in Section 7.2.2 of the text. The reader may choose the structure and values of the coefficients of the model.

5. MCMC simulation

    (a) Adapt the code for Metropolis–Hastings so that the function has informative normal priors on the intercept and the coefficient of *hmo*.

    (b) Adapt the code for Metropolis–Hastings so that the function has an informative normal prior on the intercept parameter and a beta prior on the slope.

    (c) Adapt the code for Metropolis–Hastings by adding another predictor from the `medpar` data to the model. Give it a reasonable prior.

    (d) Amend the code for Metropolis–Hastings so that it becomes a Bayesian normal model with two normal priors.

    (e) Amend the code for Metropolis–Hastings so that it becomes a Bayesian logistic model with a normal prior for the intercept and beta prior on `hmo`.

(f) Wrap the Metropolis–Hastings in a model-fitting function similar to the `ml_glm` series in the earlier chapters.

Bibliography

Aitkin, M., Francis, B., Hinde, J., Darnell, R., 2010. Statistical Modelling in R. Oxford University Press.

Akaike, H., 1973. Information theory and an extension of the maximum likelihood principle. In: Petrox, B. N., Caski, F. (Eds.), Proceedings of the Second International Symposium on Information Theory. Akademiai Kiado., pp. 267–281.

Allison, P. D., Waterman, R., 2002. Fixed-effects negative binomial regression models. Unpublished manuscript.

Baum, L. E., Petrie, T., Soules, G., Weiss, N., 1970. A maximization technique occurring in the statistical analysis of probabilistic functions of Markov chains. The Annals of Mathematical Statistics 41 (1), 164–171.

Becker, R. A., Chambers, J. M., Wilks, A. R., 1988. The New S Language: A Programming Environment for Data Analysis and Graphics. Wadworth & Brooks/Cole, Pacific Grove, CA.

Bengtsson, H., March 2003. The R.oo package — object-oriented programming with references using standard R code. In: Hornik, K., Leisch, F., Zeileis, A. (Eds.), Proceedings of the 3rd International Workshop on Distributed Statistical Computing (DSC 2003). Vienna, Austria.

Cameron, A. C., Trivedi, P. K., 1998. Regression Analysis of Count Data. Cambridge University Press, Cambridge, UK.

Casella, G., Berger, R. L., 1990. Statistical Inference. Duxbury Press, Belmont, CA.

Chambers, J. M., 1992a. Classes and Methods: Object-Oriented Programming in S. In: Statistical Models in S, Eds John M. Chambers and Trevor J. Hastie. Wadsworth & Brooks/Cole, Pacific Grove, CA.

Chambers, J. M., 1992b. Linear models. In: Statistical Models in S, Eds John M. Chambers and Trevor J. Hastie. Wadsworth & Brooks/Cole, Pacific Grove, CA.

Chambers, J. M., 1998. Programming with Data: A Guide to the S Language. Springer, New York, NY.

Chambers, J. M., 2008. Software for Data Analysis: Programming with R. Springer, New York, NY.

Chambers, J. M., Hastie, T. J., 1992. Statistical models. In: Statistical Models in S, Eds John M. Chambers and Trevor J. Hastie. Wadsworth & Brooks/Cole, Pacific Grove, CA.

Chan, T., Golub, G., LeVeque, R., 1983. Algorithms for computing the sample variance: Analysis and recommendations. American Statistician 37 (3), 242–247.

Craig, I. D., 2007. Object-Oriented Programming Languages: Interpretation. Springer, London.

DasGupta, A., 2008. Asymptotic Theory of Statistics and Probability. Springer, New York, NY.

Davison, A. C., Hinkley, D. V., 1997. Bootstrap Methods and their Application. Cambridge University Press, Cambridge, UK.

Demidenko, E., 2004. Mixed Models: Theory and Applications. John Wiley & Sons, Inc., New York, NY.

Dempster, A. P., Laird, N. M., Rubin, D. B., 1977. Maximum-likelihood from incomplete data via the EM algorithm. Journal of the Royal Statistical Society, Series B 39 (1), 1–38.

Doll, R., Hill, A. B., 1966. Epidemiological Approaches to the Study of Cancer and Other Chronic Diseases. Vol. 19. National Cancer Institute Monograph, Oxford, Ch. Mortality of British doctors in relation to smoking; observations on coronary thrombosis, pp. 204–268.

Dongarra, J. J., Moler, C. B., Bunch, J. R., Stewart, G., 1979. LINPACK Users' Guide. Society for Industrial and Applied Mathematics, Philadelphia, PA, USA.

Draper, N. R., Smith, H., 1998. Applied Regression Analysis, 3rd Edition. John Wiley & Sons, New York, NY.

Efron, B., 1979. Bootstrap methods: Another look at the jackknife. Annals of Statistics 7, 1–26.

Fisher, R. A., 1925. Theory of statistical estimation. Proceedings of the Cambridge Philosophical Society 22, 700–725.

Fitzmaurice, G. M., Laird, N. M., Ware, J. H., 2004. Applied Longitudinal Analysis. Wiley, New York, NY.

Frees, E. W., 2004. Longitudinal and Panel Data. Cambridge University Press.

Gelman, A., Hill, J., 2007. Data Analysis Using Regression and Multi-level/Hierarchical Models. Cambridge University Press.

Geman, S., Geman, D., 1984. Stochastic relaxation, gibbs distributions, and the bayesian restoration of images. IEEE Transactions on Pattern Analysis and Machine Intelligence 6, 721–741.

Gill, J., 2009. Bayesian Methods. Chapman & Hall/CRC, Boca Raton, FL.

Golub, G. H., Van Loan, C. F., 1996. Matrix Computations, 3rd Edition. Johns Hopkins University Press, Baltimore, MD.

Greene, W. H., 2012. Limdep Reference Manual, Version 10. New York, NY.

Hardin, J. W., Hilbe, J. M., 2002. Generalized Estimating Equations. Chapman & Hall/CRC, Boca Raton, FL.

Hardin, J. W., Hilbe, J. M., 2007. Generalized Linear Models and Extensions, 2nd Edition. Stata Press, College Station, TX.

Hardin, J. W., Hilbe, J. M., 2012. Generalized Estimating Equations, 2nd Edition. Chapman & Hall/CRC, Boca Raton, FL.

Harrell, F. E., 2001. Regression Modeling Strategies: With Applications to Linear Models, Logistic Regression and Survival Analysis. Springer, New York, NY.

Hartley, H. O., 1958. Maximum likelihood estimation from incomplete data. Biometrics 14 (2), 174–194.

Hastings, W. K., 1970. Monte carlo sampling methods using markov chains and their applications. Biometrika 57, 97–109.

Hilbe, J. M., 2009. Logistic Regression Models. Chapman & Hall/CRC, Boca Raton, FL.

Hilbe, J. M., 2010. Creating synthetic discrete-response regression models. Stata Technical Journal 10 (1), 104–124.

Hilbe, J. M., 2011. Negative Binomial Regression, 2nd Edition. Cambridge University Press, Cambridge, UK.

Hilbe, J. M., Linde-Zwirble, W., 1995. Random number generators. Stata Technical Bulletin 28, 20–32.

Hornik, K., 2010. The R FAQ. ISBN 3-900051-08-9.
URL http://CRAN.R-project.org/doc/FAQ/R-FAQ.html

Hsiao, C., 2003. Analysis of Panel Data. Cambridge University Press, Cambridge, UK.

Huber, P. J., 1981. Robust Statistics. John Wiley & Sons, Inc., New York, NY.

Jennrich, R. I., 1984. Comment on "Iteratively Reweighted Least Squares for Maximum Likelihood Estimation, and some Robust and Resistant Alternatives", by P. J. Green. Journal of the Royal Statistical Society, Series B (Methodological) 46 (2), 182.

Jones, O. D., Maillardet, R., Robinson, A. P., 2009. Introduction to Scientific Programming and Simulation Using R. Chapman & Hall/CRC, Boca Raton, FL.

Kates, L., Petzoldt, T., 2007. proto: Prototype object-based programming. R package version 0.3-8.
URL http://code.google.com/p/r-proto/

Lynch, S. M., 2007. Introduction to Applied Bayesian Statistics and Estimation for Social Scientists. Springer, New York, NY.

McCullagh, P., Nelder, J., 1989. Generalized Linear Models, 2nd Edition. Chapman & Hall, London.

McKendrick, A. G., 1926. Applications of mathematics to medical problems. Proceedings of the Edinbourgh Mathematical Society 44, 98–130.

McLachlan, G. J., Krishnan, T., 2008. The EM Algorithm and Extensions, 2nd Edition. Wiley, New York, NY.

Meng, X. L., van Dyk, D. A., 1997. The EM algorithm — an old folk-song sung to a fast new tune. Journal of the Royal Statistical Society, Series B 59 (3), 511–567.

Metropolis, N., Rosenbluth, A., Rosenbluth, M., Teller, A., Teller, E., 1953. Equations of state calculations by fast computing machines. Journal of Chemical Physics 21, 1087–1091.

Michalewicz, Z., Fogel, D. B., 2010. How to Solve It: Modern Heuristics, 2nd Edition. Springer–Verlag, Berlin.

Nelder, J. A., Mead, R., 1965. A simplex method for function minimization. Computer Journal 7, 308–313.

Nocedal, J., Wright, S., 2006. Numerical Optimization, 2nd Edition. Springer, New York, NY.

Pace, L., 2012. Beginning R: An Introduction to Statistical Programming. Apress.

Pawitan, Y., 2001. In All Likelihood: Statistical Modelling and Inference Using Likelihood. Clarendon Press, Oxford.

Pinheiro, J. C., Bates, D. M., 2000. Mixed-effects models in S and S-plus. Springer, New York, NY.

Polya, G., 1988. How to Solve It, 2nd Edition. Princeton University Press, Princeton, NJ.

R Development Core Team, 2012. R: A Language and Environment for Statistical Computing. R Foundation for Statistical Computing, Vienna, Austria, ISBN 3-900051-07-0.
URL http://www.R-project.org

Raftery, A., 1986. Sociological Methodology (1996). Vol. 25. Basil Blackwell, Oxford, Ch. Bayesian model selection in social research, pp. 111–163.

Robert, C., Casella, G., 2010. Introducing Monte Carlo Methods with R. Springer, New York, NY.

Schabenberger, O., Pierce, F. J., 2002. Contemporary Statistical Models for the Plant and Soil Sciences. CRC Press, Boca Raton, FL.

Schwarz, G. E., 1978. Estimating the dimension of a model. Annals of Statistics 6 (2), 461–464.

Stram, D. O., Lee, J. W., 1994. Variance components testing in the longitudinal mixed effects model. Biometrics 50 (4), 1171–1177.

Venables, W. N., Ripley, B. D., 2000. S Programming. Springer, New York, NY.

Venables, W. N., Ripley, B. D., 2010. Modern Applied Statistics in S, 4th Edition. Springer, New York, NY.

Weisberg, S., 2005. Applied Linear Regression, 3rd Edition. Wiley–Interscience, New York, NY.

Wickham, H., 2009. ggplot2: Elegant Graphics for Data Analysis. Springer, New York, NY.

Winkelmann, R., 2008. Econometric Analysis of Count Data. Springer, Berlin.

Wood, S., 2006. Generalized Additive Models. An Introduction with R. Chapman & Hall/CRC, Boca Raton, FL.

Zuur, A., Savellev, A. A., Leno, E. N., 2012. Zero-Inflated Models and Generalized Linear Mixed Models with R. Highland Statistics, Newburgh, UK.

Index